SpringerBriefs in Finance

SpringerBriefs present concise summaries of cutting-edge research and practical applications across a wide spectrum of fields. Featuring compact volumes of 50 to 125 pages, the series covers a range of content from professional to academic. Typical topics might include: a timely report of state-of-the art analytical techniques, a bridge between new research results, as published in journal articles, and a contextual literature review, a snapshot of a hot or emerging topic, an in-depth case study or clinical example, and a presentation of core concepts that students must understand in order to make independent contributions. SpringerBriefs in Finance showcase emerging theory, empirical research, and practical application in corporate finance, banking, financial management, behavioral finance, financial markets, social and entrepreneurial finance, microfinance, and related fields, from a global author community. Briefs are characterized by fast, global electronic dissemination, standard publishing contracts, standardized manuscript preparation and formatting guidelines, and expedited production schedules.

Antonella Francesca Cicchiello •
Salvatore Perdichizzi

The Role of Green Bonds in Finance

Towards a Sustainable Future

Springer

Antonella Francesca Cicchiello
IAE Business School
Buenos Aires, Argentina

Salvatore Perdichizzi
Department of Economics and Management
"Marco Fanno"
University of Padova
Padova, Italy

ISSN 2193-1720 ISSN 2193-1739 (electronic)
SpringerBriefs in Finance
ISBN 978-3-032-14839-1 ISBN 978-3-032-14840-7 (eBook)
https://doi.org/10.1007/978-3-032-14840-7

This Springer imprint is published by the registered company Springer Nature Switzerland AG
The registered company address is: Gewerbestrasse 11, 6330 Cham, Switzerland

This book is dedicated to all those who believe that building a sustainable future is both urgent and possible.

Antonella dedicates this work to her son, Andrea, whose presence fills her days with hope, joy, and purpose.

Salvatore dedicates it to his wife, Elena, for her unwavering support, encouragement, and belief in this journey.

Preface

The climate emergency is forcing capital markets to redeploy finance at unprecedented speed and scale. At that crossroads of investment and environmental policy, *green bonds* have moved fastest from niche experiment to mainstream tool. Global issuance now exceeds USD 1.6 trillion, yet key questions persist: Do labelled bonds still enjoy a cost-of-capital edge? Can voluntary standards curb green-washing? What policy architecture turns labels into measurable climate impact?

The Role of Green Bonds in Financing a Sustainable Future tackles those questions through theory, regulation, and new data, with Europe as its primary laboratory. The book serves researchers, regulators, practitioners, and graduate students who need a rigorous but accessible map of the evolving green-bond landscape.

The argument unfolds in four steps:

Chapter 1: Green Finance in Europe Surveys the full toolkit—green bonds, loans, SLBs, impact bonds, carbon markets, blue finance, crowdfunding—and flags the dual threats of data fragmentation and green-washing.

Chapter 2: Green Bonds in Practice Charts the rise of labelled debt, dissects EU milestones (Action Plan, Taxonomy, EuGB Regulation), reviews issuance drivers, and presents a tranche-level portrait of the European market (2014–2021) by country, sector, maturity, and coupon structure.

Chapter 3: Green Versus Brown Pricing Revisits the theory of the "greenium" and tests it on 3326 euro-denominated tranches (2014–2024). Once rating, sector, size, and timing are controlled for, the average greenium vanishes; any discount is confined to 2017–2019. Scarcity and disclosure credibility emerge as the decisive drivers.

Chapter 4: Conclusions and Future Research Summarises the role of green bonds in sustainable finance and highlights the importance of regulation, transparency, and trust. It also explores how AI enhances green finance through improved impact tracking and risk management, while emphasising the need for ethical, interdisciplinary approaches to achieve long-term sustainability.

Taken together, the chapters argue that green bonds will achieve their full potential only under a high-credibility equilibrium that couples stringent standards with transparent impact reporting and supportive monetary tools. We hope the book equips readers to drive both the debate and the market toward that outcome.

Buenos Aires, Argentina — Antonella Francesca Cicchiello
Padova, Italy — Salvatore Perdichizzi

Contents

About the Authors

Antonella Francesca Cicchiello is Associate Professor of Finance at the IAE Business School – Pilar (Argentina). Her teaching and research activities focus on entrepreneurial finance, FinTech, and corporate finance. She also takes an interest in Corporate Social Responsibility (CSR), green finance, and sustainable finance. She has published extensively in both academic and practitioner journals, including the *European Journal of Finance*, *Energy Economics*, *International Review of Financial Analysis*, *Finance Research Letters*, *Journal of Business & Industrial Marketing*, *Journal of International Financial Management & Accounting*, *Journal of Emerging Market Finance*, and the *International Journal of Emerging Markets*. Additionally, she serves on the editorial boards of the *International Review of Environmental and Resource Economics*, the *Journal of the Knowledge Economy*, the *European Journal of International Management*, the *International Journal of Applied Management Science*, the *World Review of Entrepreneurship, Management and Sustainable Development*, and *Cogent Economics & Finance*. She is also a Research Associate at the Banks – Banking Research School – Centre of Piacenza. She holds a PhD in Management (track in finance) from the University of Naples Federico II – Naples (Italy), in partnership with the University of Naples Parthenope – Naples (Italy).

Salvatore Perdichizzi is Associate Professor of Finance in the Department of Economics and Management "Marco Fanno" at the University of Padua. He is also a Research Fellow at SCANCOR, Stanford University, and serves as an Associate Editor of *Finance Research Letters*. His research spans monetary policy transmission, empirical banking, green and sustainable finance, and credit- and sovereign debt dynamics. He has published in leading journals such as the *Journal of Banking & Finance*, *British Accounting Review*, *Energy Economics*, *Industrial and Corporate Change*, *International Review of Financial Analysis*, *Economics Letters*, and *Finance Research Letters*. He is a Research Associate and member of the steering committee of the Yunus Social Business Centre (Bologna). He was a visiting research fellow at Stanford University, Bayes Business School, and the University of Exeter. He earned his PhD in Economics from the joint doctoral programme of the Catholic University of the Sacred Heart and the University of Milan-Bicocca.

Chapter 1
Introduction to Green Finance and Sustainable Financing

Abstract This chapter maps the European landscape of green finance, situating green bonds within a broader architecture that spans green and sustainability-linked loans and bonds (SLB/SLL), blue finance (blue bonds and loans), environmental impact bonds, green credit lines, green equity, and retail channels such as crowdfunding. It synthesises the policy drivers behind Europe's leadership—taxonomy-based classification and disclosure regimes (e.g. SFDR) under the Green Deal—while documenting rapid market scaling alongside vulnerabilities: greenwashing incentives, data fragmentation, and cross-country heterogeneity. The chapter distinguishes use-of-proceeds, performance-linked, and outcome-based designs and explains how eligibility criteria, external reviews, and ex post impact reporting (MRV) support investor trust and allocative efficiency. It clarifies the role of carbon pricing and markets—cap-and-trade (EU ETS) versus offsets—and the contribution of carbon-finance derivatives to hedging and liquidity. Finally, it offers a risk map (standardisation gaps, verification quality, reputational and regulatory exposure) and a policy agenda centred on taxonomy alignment, comparable metrics, and robust assurance. A recent strand of evidence from pan-European credit registers shows that public loan guarantees (PGL) introduced during COVID-19 tilted credit supply towards low-emission firms by relaxing monitoring constraints, highlighting both the information-production cost of green lending and the importance of programme design for portfolio greening (Buchetti et al., Loan guarantee and portfolio greening: evidence from European credit registers, 2025). The contribution is a system view: Instruments are effective when embedded in credible standards, data governance, and incentive-compatible programmes that connect capital to measurable environmental outcomes.

Keywords Green finance · Public loan guarantees (PGL) · Green bonds · Sustainability-linked bonds and loans · EU taxonomy · SFDR · Carbon markets (EU ETS · offsets) · Blue finance · Crowdfunding · Green equity · Monitoring–reporting–verification (MRV) · Greenwashing

A. F. Cicchiello, S. Perdichizzi, *The Role of Green Bonds in Finance*, SpringerBriefs in Finance, https://doi.org/10.1007/978-3-032-14840-7_1

1.1 Overview of Green Finance Development in Europe

Green finance, or the mobilisation of financial resources for sustainable development projects and initiatives, has become a key pillar of the global transition to a low-carbon and environmentally sustainable economy (Akomea-Frimpong et al. 2022). The roots of green finance trace back to ethical investment movements and early corporate social responsibility (CSR) efforts of the late 20th. Over time, the concept expanded to embrace Environmental, Social, and Governance (ESG) criteria, aligning closely with the sustainable development goals (SDGs), adopted by the United Nations in 2015. As Migliorelli and Dessertine (2019) argue, the integration of sustainability into financial markets reflects the urgency of the climate crisis, the growing significance of ESG considerations, as well as the increasing demand for sustainable investment products. The European Union (EU), through a combination of regulatory initiatives, market innovations, and institutional collaboration, has positioned itself as a global leader in green finance. The EU Green Deal, introduced in 2019, articulates a vision for achieving climate neutrality by 2050. A key component of this strategy is the Sustainable Finance Action Plan, which aims to channel capital towards sustainable investments. Foundational tools such as the EU Taxonomy, a classification system for sustainable economic activities, and the Sustainable Finance Disclosure Regulation (SFDR), which enhances ESG transparency, have played a key role in achieving this leadership position. In parallel, European financial markets have increasingly embraced green finance, recognising its strategic importance for long-term value creation (Perdichizzi et al. 2024) and alignment with global sustainability goals. The integration of green finance within capital markets has become evident through the rapid growth of green finance instruments, such as green bonds, sustainability-linked loans, and ESG-focused investment funds (Cardillo et al. 2024).

Recent micro-evidence from pan-European credit registers further nuances this system view: During the COVID-19 guarantee programmes, banks used public loan guarantees (PGL) to expand credit to low-emission firms, particularly in sectors less affected by the pandemic shock, and were more likely to open new relationships with such firms; at the same time, internal risk monitoring (probability-of-default updates) was relaxed under guarantees and became less predictive for low-emission borrowers—consistent with higher information-production costs in green lending and with the programme's design shifting downside risk to the public sector (Buchetti et al. 2025). This evidence complements the EU's supervisory push on climate risk management by clarifying how incentives and operational frictions shape the pace of portfolio greening at banks.

Despite significant progress, green finance development in Europe faces several challenges. One key issue is the lack of standardised metrics and methodologies for assessing sustainability performance. This lack of standardisation undermines comparability and increases the risk of greenwashing, where products are labelled as sustainable without demonstrable adherence to recognised sustainability criteria (Di Tommaso et al. 2024). Moreover, disparities in market maturity, regulatory frameworks, and investor awareness across European Member States hinder the development of a cohesive and inclusive green finance ecosystem (Afzal et al. 2022). Bridging these gaps is essential for realising the full potential of green finance and ensuring that the green transition is both inclusive and equitable.

1.2 Green Finance Instruments

Green finance instruments are specialised financial tools designed to support environmentally sustainable economic activities (Babic 2024). They mobilise capital towards projects that generate measurable environmental benefits, such as climate change mitigation, clean energy, pollution control, biodiversity conservation, and sustainable land and ocean use. As global awareness of climate risks intensifies and regulatory frameworks evolve (Cicchiello et al. 2023c), the financial sector has responded with a variety of instruments that embed sustainability objectives into lending, investment, and risk-sharing mechanisms (Monk and Perkins 2020).

These instruments differ in structure, investor type, repayment terms, and sectoral focus. Some, like green bonds and green loans, require that proceeds be exclusively allocated to eligible green projects. Others, such as sustainability-linked loans or bonds, tie financial conditions (e.g. interest rates) to the borrower's achievement of specific environmental or social performance targets (Cicchiello et al. 2022a). Additionally, innovative instruments such as blue bonds, impact bonds, green crowdfunding, and carbon pricing tools are expanding the scope of green finance to address complex ecological challenges, particularly in marine and public-sector contexts (Choudhary et al. 2021; Carè et al. 2025).

Together, these instruments form a growing ecosystem of sustainable finance solutions aimed at redirecting financial flows in line with international commitments like the Paris Agreement and the UN Sustainable Development Goals (SDGs), especially SDG 13 (Climate Action) and SDG 14 (Life Below Water) (Iacobuţă et al. 2022).

Table 1.1 presents a structured overview of the most relevant green and sustainable finance instruments currently in use. It categorises each instrument by providing a concise explanation of the financial tool and its key features, the type (i.e. debt, equity, or mixed), the sectors where the instrument is most commonly applied, and real-world cases or prominent issuances that illustrate the application of each instrument.

The table includes traditional tools like green bonds and green loans, as well as innovative mechanisms such as blue bonds, sustainability-linked instruments, and carbon pricing mechanisms. It also highlights the evolving role of retail-based financing (e.g. green crowdfunding) and outcome-based models (e.g. environmental impact bonds and development impact bonds).

The following sub-paragraphs describe how each instrument operates, what environmental goals it targets, and how it aligns with broader sustainability strategies.

Table 1.1 Green and sustainable finance instruments

Instrument	Type	Definition	Primary use/sector	Examples
Green bonds	Debt	Fixed-income securities used to finance environmental projects	Renewable energy, energy efficiency, transport	France green OAT; apple green bond
Green loans	Debt	Loans with proceeds restricted to green uses	Buildings, infrastructure, energy efficiency	IKEA green loan for sustainable buildings
Sustainability-linked bonds (SLBs)	Debt (KPI-tied)	Bonds where interest rates adjust based on issuer's sustainability performance	Corporate/ sovereign sustainability targets (flexible use)	Enel SLB linked to GHG reduction
Sustainability-linked loans (SLLs)	Debt	Loans tied to borrower's ESG performance via KPIs	Corporates (multi-sector)	Danone's SLL with carbon and water targets
Environmental impact bonds (EIBs)	Debt (outcome-based)	Outcome-based financing in which repayments depend on realised environmental impact	Urban resilience, storm water, wetlands	Washington, D.C. green-infrastructure EIB
Green credit lines	Credit line	DFI or central-bank credit lines dedicated to green transitions.	Public utilities, SMEs, energy access	EBRD's Green Economy Financing Facility (GEFF)
Green equity	Equity	Equity strategies prioritising environmental, social, and governance criteria	All sectors; listed equities, ETFs, mutual funds	MSCI ESG Index; BlackRock ESG portfolios
Carbon markets/derivative products	Incentive	Market tools that assign a cost to carbon emissions (e.g. taxes, cap-and-trade)	Energy, industry, aviation, agriculture	EU ETS; California Cap-and-Trade
Green crowdfunding	Mixed	Retail-based financing of environmental projects via digital platforms	Solar, sustainable agriculture, clean energy	One Planet Crowd; Abundance (UK); Wiwin (DE)
Blue bonds	Debt	Bonds dedicated to financing marine and coastal protection	Sustainable fisheries, coral reefs, ocean resilience	Seychelles Blue Bond (2018); Belize Blue Bond (2021)
Blue loans	Debt	Loans funding ocean-based sustainable activities	Clean ports, marine renewables, aquaculture	IFC Blue Loan to Indorama Ventures

Source: Authors elaboration

1.2.1 Green Bonds and Green Loans

Green bonds and green loans are two key instruments in the rapidly evolving sustainable finance landscape, enabling the flow of capital into projects and activities that deliver measurable environmental and social benefits (Flammer 2020). Green bonds are debt securities where proceeds are exclusively dedicated to financing or refinancing eligible green projects (Cicchiello et al. 2022b). They represent the largest and most prominent sustainable financial instruments globally. Since the first "Climate Awareness Bond" was issued by the European Investment Bank (EIB) in 2007, the global green bond market has expanded rapidly, with cumulative green bond issuance reaching $1.4 trillion between 2012 and 2021 (Babic 2024).

Green bonds are classified based on the entity issuing them, the structure of the bond, and the use of proceeds (ICMA 2024). In terms of issuer, green bonds can be issued by sovereign entities (i.e. national governments), sub-sovereign entities (i.e. municipalities, regional authorities), supranational institutions (e.g. European Investment Bank, World Bank), corporations, and financial institutions. Governments and financial institutions remain the largest issuer categories. According to the Climate Bonds Initiative report (Climate Bonds Initiative 2025) and latest EU market data, France remains the world's largest sovereign issuer of green bonds, followed by Germany, with Italy and the United Kingdom among the top five. While the United States remains a leading green bond market overall, its issuance is primarily through municipal (sub-sovereign) green bonds, as the federal government has not yet issued a labelled sovereign green bond. The second form of classification relates to the debt structure, as recognised in the Green Bond Principles. Green bonds may be structured as Standard Green Use-of-Proceeds Bonds (typically plain-vanilla senior unsecured bonds), Green Revenue Bonds (supported by revenues from green projects), Green Project Bonds (secured by project-level cash flows), or Secured Green Bonds, including covered bonds or asset-backed securities (ABS) linked to pools of green assets. Finally, the third form of classification concerns the use of proceeds, which lies at the heart of the Green Bond Principles, as it is what fundamentally distinguishes green bonds from conventional bonds. Green bond proceeds must be exclusively dedicated to projects that generate measurable environmental benefits. The GBP define several core eligible categories, including: renewable energy, energy efficiency, pollution prevention and control, sustainable management of natural resources and land use, biodiversity conservation, clean transportation, sustainable water management, climate change adaptation, and green buildings (Climate Bonds Initiative 2025). Issuers are required to explicitly state the intended use of proceeds and report regularly on project allocation and environmental impact.

Green loans, while less extensively documented than green bonds, are a fast-growing segment of the sustainable finance market. They can be defined as any type of loan instrument made available exclusively to finance or refinance eligible green projects (Giraudet et al. 2021). Green loans are governed by the Green Loan Principles, published by the Loan Market Association, which establish four core components: (i) use of proceeds; (ii) process for project evaluation and selection; (iii) management of proceeds; and (iv) reporting. Loan proceeds must be allocated exclusively to eligible green projects delivering measurable environmental benefits, across categories such as renewable energy, energy efficiency, clean transport, biodiversity conservation, and green buildings. Borrowers must clearly communicate project selection processes, manage proceeds transparently, and provide regular updates on the use of funds and project impacts. While external reviews are recommended, self-certification is permitted when borrowers demonstrate appropriate internal expertise.

1.2.2 Sustainability-Linked Bonds and Sustainability-Linked Loans (SLBs and SLLs)

Sustainability-linked finance refers to loans and bonds whose interest terms are linked to the issuer's progress on sustainability performance criteria. Rather than requiring funds to be directed towards predefined green or social projects, as in use-of-proceeds instruments, sustainability-linked instruments allow for unrestricted use of proceeds (Ehlers et al. 2021). Their distinguishing feature lies in aligning financial costs, such as interest rates, with measurable improvements in environmental, social, or governance (ESG) performance.

Sustainability-linked bonds (SLBs) are an innovative type of bond instrument in which the financial and/or structural characteristics (typically the coupon rate) vary based on the issuer ability to achieve predefined sustainability performance targets (SPTs) (Mishra et al. 2023). These targets are measured through predefined key performance indicators (KPIs) related to environmental, social, and governance (ESG) factors.

According to the ICMA's sustainability-linked bond principles (ICMA 2024), the KPIs should be:

1. Highly relevant and strategically significant to the issuer's core business or sustainability policies
2. Aligned with the issuer's sustainability strategy and focused on its most material strategic priorities, potentially supported by disclosures or transition plans in hard-to-abate sectors
3. Measurable or quantifiable on a consistent methodological basis
4. Externally verifiable
5. Able to be benchmarked

As general-purpose bonds, sustainability-linked bonds allow the issuer to set broader sustainability targets, rather than being tied to financing specific environmental or social projects as in the case with green and sustainability bonds (Kölbel and Lambillon 2022). In essence, the issuer commits to enhance its overall sustainability performance, regardless of whether the funds are used for specific green-labelled projects. If the issuer does not meet the agreed sustainability performance targets (SPTs), the bond may trigger a step-up mechanism, such as an increase in the interest rate or other adjustments to its financial terms.

The first sustainability-linked bond (SLB) was issued by Enel S.p.A., a multinational energy company based in Italy, in September 2019 (Feldhütter et al. 2024). Since then, cumulative SLB issuance globally surpassed USD 230 billion, showing their potential to promote the transition to a more sustainable economy (Vulturius et al. 2024). According to data from Environmental Finance, SLBs now represent approximately 17–20% of the GSSS (Green, Social, Sustainability, and SLB) bond market.

While sustainability-linked bonds (SLBs) offer a flexible and performance-driven approach to sustainable finance, they also carry distinct risks and challenges for both issuers and investors. These risks range from the potential for greenwashing, particularly when issuers select non-material or easily achievable KPIs, to financial risks related to the step-up mechanisms which adjust the bond's interest rate if the issuer fails to meet its targets (Giráldez and Fontana 2022). These mechanisms introduce uncertainty into expected returns and may increase the cost of capital, particularly for firms operating in sectors with volatile ESG metrics (Broadway 2022). Lastly, the SLB market, while expanding, still lacks the depth and liquidity of green bonds, potentially limiting diversification options, especially in emerging economies. Addressing these risks will be essential for the long-term viability and impact of SLBs as a central instrument in the sustainable finance landscape.

Similarly to SLBs, sustainability-linked loans (SLLs) are loan agreements where the terms, typically the interest rate, are directly tied to the borrower's achievement of predetermined sustainability performance targets (SPTs). Unlike green loans, which require the loan proceeds to be used exclusively for environmental or social projects, SLLs can be used for general corporate purposes, offering greater flexibility (Du et al. 2023). What sets SLLs apart is that the borrower commits to improving performance on material environmental, social, or governance (ESG) metrics, and the financial terms of the loan adjust based on success or failure in meeting those targets.

1.2.3 Environmental Impact Bonds (EIBs)

Environmental impact bonds (EIBs) are a results-based financing instrument specifically designed to fund projects with measurable environmental outcomes (Trotta 2024). They mobilise upfront private capital and link the issuer's repayment profile to verified ecological performance—such as reductions in combined sewer overflows, stormwater retention, nutrient load abatement, acreage of restored wetlands, or persistence of tree-canopy gains over time. In contrast with traditional use-of-proceeds green bonds, where monitoring focuses on allocation and eligibility, EIBs embed performance risk in the security's cash-flow schedule, rewarding successful delivery and, symmetrically, reducing payments when outcomes fall short. The approach creates disciplined incentives along the project lifecycle, raises the salience of monitoring–reporting–verification (MRV), and can accelerate organisational learning for public sponsors.

The first widely cited EIB was launched by DC Water in 2016 to finance green-infrastructure retrofits aimed at mitigating urban flooding and reducing combined sewer overflows. The bond was privately placed and featured a performance-contingent payment at a predefined evaluation date linked to measured stormwater capture. Subsequent transactions have adapted the template to public offerings and larger programmes. In 2019, Atlanta's Department of Watershed Management issued a publicly offered EIB to fund multiple green-infrastructure sites in underserved neighbourhoods. In 2021, the Buffalo Sewer Authority closed what was then the largest EIB for stormwater mitigation as part of its Rain Check 2.0 initiative, combining a conventional debt structure with an outcome-based component tied to verified reductions in overflow risk. These cases illustrate that the EIB is not a niche pilot but a repeatable contractual architecture that public utilities can adapt to local hydrology, data systems, and regulatory constraints.

From a financial-economics perspective, the hallmark of an EIB is the mapping from a verified environmental outcome to a payment schedule. Let E denote the realised outcome (for example, avoided overflow events per year) and E_0 the agreed baseline. A stylised schedule can be represented as

$$\text{Payment} = \alpha + \beta\,(E - E_0) \quad \text{subject to caps/floors and minimum-performance thresholds,}$$

where α captures any fixed service component and β is the marginal compensation per unit of impact. In practice, public sponsors often implement this logic with a small number of discrete tiers—for instance, a lower payment when performance is below a threshold, a base payment within an expected band, and a success payment if outcomes exceed target. Caps on both performance and money bound fiscal exposure; floors protect investors from idiosyncratic shortfalls unrelated to delivery (for example, exogenous weather anomalies beyond the contract's force-majeure window). The schedule must be complemented by a clear treatment of measurement error and uncertainty. Because the signal E is estimated from sensors and models, contracts frequently include tolerance bands around thresholds and require uncertainty quantification in the MRV protocol so that random noise neither triggers windfall gains nor penalises genuine performance.

The credibility of the counterfactual is central. Baselines can be constructed from historical series corrected for rainfall regimes, from matched-control areas under a difference-in-differences design, or, where feasible, from staggered roll-out that approximates experimental variation. Each option trades econometric rigour against administrative complexity. Where outcomes exhibit long lags—for example, nutrient load reductions at downstream monitoring stations—interim milestones based on well-validated leading indicators can support partial payments and thus reduce the carry cost for investors without diluting accountability for end-state impacts. Contracts should also address leakage (displacement of runoff or pollution to untreated areas), rebound (behavioural responses that offset gains), and persistence (degradation of benefits through time), either through explicit adjustments to E or through clawback provisions tied to post-completion performance.

EIBs require a robust MRV stack anchored in transparent data governance. For stormwater and urban heat projects, cities increasingly deploy flow metres, IoT rain gauges, and remote sensing (LIDAR for canopy, SAR for wetland hydrology) integrated with hydrologic-hydraulic models pre-specified in the term sheet or its technical annexes. Good practice involves archiving raw data, model versions, and parameter files with time-stamped hashes and third-party audit rights, so that verification is reproducible ex post and learning is cumulative across vintages. Independent evaluators should be engaged early to co-design the protocol, test the instrumentation plan, and run pre-mortems on common failure modes (for example, sensor drift, missing observations, or boundary-condition mis-specification). Publishing a machine-readable MRV summary after each verification not only strengthens accountability for the specific deal but also lowers transaction costs for subsequent issuances by standardising templates.

Because EIB cash flows are formally contingent, the investor base tends to include actors with mandate flexibility and a stated impact objective. Philanthropic and mission-driven capital frequently provides junior protection (first-loss equity or guarantees), while insurance balance sheets and impact debt funds subscribe senior paper. The internal rate of return depends on three drivers: the distribution of outcomes under realistic hydrologic scenarios; the cadence of verification and settlement (which affects duration and reinvestment risk); and the strength of downside protection via floors or step-down mechanisms. Where verified outcomes translate into monetisable savings for the sponsor—avoided regulatory penalties for combined sewer overflows, reduced emergency response, lower pumping costs—EIBs can be integrated into "pay-as-you-save" frameworks that align debt service with realised fiscal benefits. This alignment is attractive to credit committees and rating agencies because it anchors repayment capacity in operating cash flows rather than in discretionary appropriations.

The public-finance interface deserves careful design. Outcome-contingent obligations may require explicit authorisation and multi-year appropriation language. Procurement rules must allow the sponsor to specify outputs and verification standards rather than inputs, so that bidders can innovate on delivery while competing on price per unit of outcome. Auditors and treasurers need clear policies for recognising contingent liabilities and for ring-fencing success-payment reserves to avoid inter-temporal budget gaming. Where state aid or intergovernmental transfers are involved, compatibility with eligibility rules must be checked ex ante. These institutional considerations are not peripheral: They are often the difference between a one-off pilot and a scalable programme.

The strengths of the EIB model—allocative efficiency, dynamic discipline, and catalytic leverage—are matched by predictable pitfalls. If the metric is too narrow, implementers may optimise the proxy rather than the underlying welfare objective.

If the metric is too complex or data-hungry, verification costs can consume the very efficiency gains the instrument intends to deliver. If siting decisions chase the cheapest units of impact, distributional inequities can widen, especially in cities where flood risk and heat exposure correlate with socioeconomic disadvantage. These risks can be mitigated by complementing the primary outcome with a small set of co-benefit indicators disclosed—but not necessarily remunerated—in the MRV report; by publishing the siting algorithm and embedding equity constraints ex ante; and by stress-testing the payment schedule against plausible measurement errors and weather regimes. In other words, governance of information—clarity, auditability, and parsimony—is a first-order design choice.

EIBs should also be situated relative to neighbouring instruments. When outcomes are observable over the programme horizon and identification is credible, EIBs can deliver higher impact per euro than use-of-proceeds bonds because they purchase verified services rather than planned inputs. When outcomes are diffuse or lagged, a classic green bond with strong allocation and impact reporting may dominate on simplicity and cost. Sustainability-linked bonds shift incentives to the issuer level via coupon step-ups tied to corporate KPIs; they can coexist with project-level EIBs in the same capital structure, with the former steering trajectory and the latter contracting for specific environmental services. Related pay-for-success models, such as forest resilience bonds used for wildfire-risk reduction and watershed protection, demonstrate that outcome-based repayment can be adapted beyond urban water to ecosystems where benefits accrue to multiple beneficiaries over time. The unifying economic logic is to tie capital remuneration to delivered externality reduction under a verification regime that the market trusts.

A brief numerical example clarifies fiscal risk. Suppose a utility sets a baseline of 100 overflow events per year and a target of a 30-event reduction within three years. It agrees to pay a fixed 1 million availability component upon verified completion and 200,000 per avoided event, capped at the target. If verification confirms 30 events avoided, the outcome payment totals 7 million; if performance reaches only 15, the total is 4 million. With a five-year tenor and semi-annual settlement, expected investor returns are then a function of the weather-adjusted distribution of outcomes and the agreed caps and floors. From the sponsor's perspective, the schedule bounds exposure while purchasing precisely the service it values—overflow avoidance—at a price revealed through competition.

Finally, scalability hinges on standardisation. Term-sheet templates, interoperable MRV protocols, and pooled vehicles that aggregate smaller municipalities can compress transaction costs without sacrificing integrity. A modest upfront investment in open data schemas and registries that record contracts, protocols, and verification summaries would accelerate secondary learning and, by reducing diligence costs, expand the feasible universe of issuers. In that equilibrium, EIBs become a durable complement to traditional labelled debt: They are deployed where outcomes can be credibly measured within policy-relevant horizons, while use-of-proceeds bonds and sustainability-linked instruments serve programmes and issuer-level transitions less amenable to tight outcome contracting.

1.2.4 Green Credit Lines

Green credit lines are specialised wholesale facilities provided by multilateral development banks (MDBs), development finance institutions (DFIs), or national promotional banks to local intermediaries that on-lend to households, firms, and municipalities for environmentally eligible investments. Unlike conventional credit programmes, they bundle earmarked term funding with a standardised eligibility framework and dedicated technical assistance. The design addresses three frictions that keep profitable decarbonisation projects from being financed at scale: information and appraisal costs for technologically heterogeneous, small-ticket investments; perceived performance risk, amplified by imperfect measurement; and the wedge between private and social returns created by unpriced environmental externalities. In the language of intermediation theory, the facility lowers the cost of delegated monitoring and screening (Diamond 1984), relaxes credit rationing induced by asymmetric information (Stiglitz and Weiss 1981), and eases funding-liquidity constraints that impede the translation of loanable funds into real investment (Holmström and Tirole 1997). Experience from the EBRD's Green Economy Financing Facility and AFD's SUNREF initiative shows that, when these elements are coherently integrated, local banks expand the feasible frontier of bankable energy- and resource-efficiency projects while retaining credit discipline.

The microeconomics of the instrument can be expressed compactly. Consider project j with investment outlay K_j yielding private cash flows $CF_{j,t}$ and environmental services ΔE_j (e.g. energy saved, emissions avoided). A local bank faces an appraisal cost $k(\theta_j)$ increasing in the project's technical complexity θ_j, a funding cost r_M on market liabilities, and a capital charge implicit in its hurdle rate. In a conventional loan, the bank lends if

$$\mathbb{E}\left[\sum_t \frac{CF_{j,t}}{(1+r_M)^t}\right] - K_j - k(\theta_j) \ \geq \ 0,$$

after collateral and covenant adjustments. A green credit line replaces the marginal funding cost r_M with facility funding $r_L \leq r_M$, supplies advisory $\tau(\theta_j)$ that lowers appraisal cost $(k(\theta_j) - \tau(\theta_j))$, and may include a contingent incentive $R(\Delta E_j)$ disbursed at verified completion. The local condition becomes

$$\mathbb{E}\left[\sum_t \frac{CF_{j,t}}{(1+r_L)^t}\right] - K_j - \left[k(\theta_j) - \tau(\theta_j)\right] + R(\Delta E_j) \ \geq \ 0.$$

Three margins move: a funding-cost margin (maturity-matched, typically longer-tenor liabilities), an information-cost margin (codified technology lists, engineering support, origination templates), and an incentive margin (completion-verified rebates or performance-based top-ups). Even when $R(\cdot)$ is modest or zero, the $(r_M \downarrow, k \downarrow)$ shift can be sufficient to cross the credit threshold for dispersed, modular projects with positive private NPV but high appraisal friction. From a social standpoint, measuring and disclosing ΔE_j narrow the wedge between private and social NPV and generate evidence that informs policy on cost-effectiveness across technologies and sectors.

The facility's credibility rests on an eligibility taxonomy and an auditable monitoring, reporting, and verification (MRV) protocol. At origination, each sub-loan is screened against technology- or outcome-based criteria and records a counterfactual baseline, investment costs, and ex ante resource-savings estimates using standard coefficients. At completion, documentation (invoices, commissioning evidence) and, where warranted, site verification anchor ex post estimates. The aggregation logic—persistence assumptions, degradation factors, and default coefficients—must be specified ex ante to keep portfolio indicators consistent and resistant to discretionary reclassification. This design balances the need to minimise transaction costs for small tickets with the requirement that environmental claims be verifiable and comparable across intermediaries and time.

Risk sharing remains with the local bank for borrower credit risk, preserving incentives for screening and monitoring; the MDB or DFI bears country and intermediary exposure on the wholesale line. Guarantees and first-loss tranches can be layered to accelerate adoption in new segments or extend maturities, but blanket protection weakens underwriting discipline and invites selection on softness. Currency mismatches require explicit treatment: Whereas the wholesale facility is foreign-currency funded and retail on-lending is in local currency, feasible designs include local-currency windows, facility-level hedging, or pricing bands that allocate residual risk transparently. A facility manual that fixes these conventions ex ante avoids ad hoc renegotiations that erode the signal value of the programme.

The role of technical assistance is not ancillary. On the supply side, credit officers learn to integrate engineering parameters into cash-flow lending, internal scorecards incorporate environmental fields, and standard credit-memo templates reduce dispersion in appraisal. On the demand side, SMEs and municipalities receive support to scope measures, obtain vendor quotes, and prepare bankable proposals that meet documentation standards. Over successive cohorts, repeated exposure reduces perceived novelty risk and makes origination faster and cheaper. This institutional learning—rather than the initial concessional margin per se—is typically the durable legacy: Participating banks establish green desks, vendor frameworks, and segment-specific risk appetites that persist beyond the facility's life.

The tension between additionality and subsidy is central to design. Investment grants, interest-rate rebates, or completion bonuses can be justified when they compensate for genuine externalities or coordination failures. Calibrated poorly, they crowd out private finance, select for low-effort projects, and dampen the disciplining role of borrower skin-in-the-game. The operational remedy is straightforward: Keep borrower co-financing meaningful; make any incentive contingent on verified completion; cap support as a fraction of eligible costs; and publish summary statistics on uptake, defaults, and measured savings to permit external scrutiny. In environments with thin green deal flow or weak vendor markets, temporary, targeted incentives can catalyse entry; as pipelines mature, subsidies should phase down, while standardisation and data continue to compress appraisal costs.

Integration with prudential and strategic constraints determines scalability. For banks, stable wholesale funding at maturities aligned with asset profiles is valuable, but not if eligibility rules collide with core systems or if compliance overhead overwhelms origination margins. Facilities scale when the taxonomy maps cleanly to national standards (or, in Europe, to the EU Taxonomy), when MRV templates are interoperable with the bank's credit workflow, and when supervisory recognition is clear. As repeat issuance builds track records on defaults and realised savings, the product migrates from a niche to business-as-usual: Pricing reflects true risk; pipelines are generated through vendor partnerships; and internal dashboards manage financial and environmental KPIs side by side.

Country context shapes sectoral emphasis without altering the underlying economics. In lower-income settings, household and micro-enterprise upgrades dominate because paybacks are short and documentation simple. In middle-income economies, commercial building retrofits, industrial process optimisation, distributed renewables, and agricultural water efficiency expand as tariff structures and regulation stabilise. Waste-management, cold-chain modernisation, and water-loss reduction have proven bankable where eligibility is precise and advisory is present. Aligning the facility's taxonomy with national climate and energy strategies, and disclosing machine-readable portfolio aggregates, allows policymakers to update sectoral priorities on observed cost-effectiveness rather than ex ante modelling alone.

Evaluation is feasible with standard empirical designs. Because eligibility rules and technology lists often create thresholds, difference-in-differences around launch and strengthening events, or design-based strategies exploiting cut-offs, can identify incremental credit and environmental effects at the margin. Bank-firm panels allow tests for crowding-out versus crowding-in at the borrower level; branch-level data can capture geographic diffusion and vendor spillovers; and survival analysis on loan performance addresses concerns about hidden risk. In programmes that introduce completion bonuses, event-study windows around verification dates provide direct tests of moral-hazard mitigation. The binding constraint is typically data governance rather than methodology: Credible evaluation requires that facility-level datasets link financial and environmental fields at the sub-loan level under clear privacy protocols.

The contrast with other labelled instruments clarifies the niche. Use-of-proceeds green bonds are efficient for large, homogeneous capex where appraisal is already internalised; they do not, by themselves, alter the bank's cost of discovering and underwriting thousands of dispersed upgrades. Sustainability-linked instruments align issuer-level incentives but are weak at contracting specific, verifiable environmental services at the borrower level. Green credit lines, by construction, intervene precisely where those other tools are least effective: granular, modular investments whose private economics are positive but whose appraisal and verification costs deter intermediation. When the bundle of long-tenor funding, standardised eligibility, and advisory is executed with proportionate MRV and prudent risk sharing, they scale bankable decarbonisation while embedding capabilities in the local financial system.

1.2.5 Green Equity

Green equity comprises equity investments that finance environmentally sustainable activities in renewables, energy efficiency, clean infrastructure, and adjacent enabling technologies (Babic 2024). Vehicles range from private equity funds and public-equity strategies to publicly backed capital channelled through development finance institutions (DFIs) and multilateral climate funds, whose mandate is catalytic rather than purely financial (Amighini et al. 2022). Post-Paris, the segment has scaled materially, yet it remains small relative to global equity markets and vulnerable to misclassification and disclosure heterogeneity. The effective scope and impact of green equity therefore hinge on the credibility of classification standards and on the reliability of environmental data and scoring methodologies (Utz 2019).

From a corporate-finance perspective, the economic content of green equity is the joint relaxation of financing and governance frictions for projects with high intangible intensity, steep learning curves, and option value. When verification is costly and payoffs arrive with long lags, internal cash flows or bank credit are weak substitutes for risk capital endowed with monitoring rights. Specialised equity reduces the shadow cost of financing constraints by underwriting experimentation and scale-up, and it raises the marginal productivity of investment by supplying governance that disciplines strategy, capital budgeting, and disclosure. Where stewardship is credible and exerted through boards and incentive contracts, the firm's policy vector—capex composition, R&D sequencing, and asset retirements—tilts towards transition-consistent choices in ways that are ultimately visible in operating metrics and valuation (Cicchiello et al. 2023c; Perdichizzi et al. 2024).

On the asset-pricing margin, expected returns on "green" equities embed both preference and risk components. Taste for environmental characteristics can compress required returns for credibly aligned issuers; by contrast, exposures to systematic transition risk—policy discontinuity, technology displacement, or demand substitution—raise premia when they are not easily hedged. Separating these channels is an identification problem that turns on measurement. Disagreement across ESG raters, unstable mappings from narrative to numeric scores, and the treatment of scope-3 emissions in complex value chains attenuate the signal-to-noise ratio and confound inference (Utz 2019; Bissoondoyal-Bheenick et al. 2024). In benchmarked mandates, index design reinforces these issues: Exclusion-based tilts alter sector exposures and can import unintended factor bets, whereas sector-neutral decarbonisation frameworks achieve large footprint reductions with limited tracking error but do not automatically reallocate capital to the highest-abatement opportunities unless explicitly linked to capacity expansion (Andersson et al. 2016;

Pedersen et al. 2021). In practice, fundamental, thesis-driven strategies anchored in technology economics, unit-level energy intensity, and time-stamped transition plans dominate naive screens on both performance attribution and impact credibility.

The relevant distinction for policy and evaluation is between portfolio impact and real-world impact. Portfolio decarbonisation is feasible by rebalancing into already low-emission firms without changing aggregate abatement. Real-world impact requires that invested equity causes additional capacity or accelerate technology substitution relative to a credible counterfactual. Two mechanisms deliver such additionality. First, primary issuances and private rounds that finance new plants, retrofits, or product redesigns expand abatement supply. Second, stewardship that conditions follow-on capital or compensation on verifiable operational milestones shifts incumbent trajectories. Both are testable with standard designs: Event windows around capital raisings and shareholder interventions, combined with panel evidence on capex, R&D, and emissions intensity, permit attribution of investor-induced changes as distinct from demand or policy shocks (Cicchiello et al. 2023c). The credibility of any impact claim rests on pre-analysis plans, explicit baselines, and ex post reporting that separates organic de-risking from investor causation.

DFIs and climate funds play a catalytic role where country, policy, or technology risks deter commercial equity. Blended structures that absorb first-loss risk or stabilise cash flows can crowd in private investors without dulling operating discipline, provided support is transparent, time-bounded, and contingent on measurable outcomes (Amighini et al. 2022). In public markets, quasi-policy investors can anchor seasoned offerings by issuers that meet robust taxonomy alignment and disclosure standards, compressing issuance costs and broadening ownership. Without tight guardrails, however, concessional layers risk subsidising incumbency rather than accelerating diffusion.

Data governance remains the binding constraint. Credible security selection and accountability require project-level emissions, boundary-consistent scope-3 estimates were material, and audit trails for avoided-emissions claims. Where mandatory regimes such as the EU Taxonomy and the Corporate Sustainability Reporting Directive apply, assurance and comparability reduce information asymmetries and, by construction, the cost of capital for issuers that internalise them (Babic 2024). Where disclosure is principles-based, investors can contract for decision-useful data by embedding transition covenants in governance and by conditioning follow-on capital on delivery of machine-readable metrics (Cicchiello et al. 2023c). In all cases, the durable test for green equity is whether one can trace an auditable line from capital deployed, through governed information and corporate policy adjustment, to realised abatement. Without that evidentiary chain, portfolio greenness is indistinguishable from relabelling; with it, equity becomes a credible instrument for reallocating risk and control towards projects that shift the production possibility frontier of decarbonisation at scale.

1.2.6 *Carbon Markets and Carbon-Finance Derivative Products*

Carbon markets and carbon pricing instruments are key tools in the fight against climate change. They are designed to internalise the environmental cost of greenhouse-gas (GHG) emissions and incentivise emission reductions across all sectors of the economy in line with Paris Agreement ambitions to limit global warming.

Carbon markets are the carbon-emissions trading market. They represent a central component of world climate change policy, aimed at reducing greenhouse-gas emissions by placing a cost on carbon dioxide through market-based instruments (Calel 2013). These markets operate under two major systems: cap-and-trade systems and carbon-offset schemes (Fankhauser and Hepburn 2010). In cap-and-trade systems, such as the EU Emissions Trading System (EU ETS), a central authority sets a cap on the total amount of GHGs that can be emitted by covered entities. This cap is divided into emission allowances, each of which gives the right to release a specific amount of CO_2 (typically one tonne). Companies are required to hold enough allowances to cover their emissions, but they can buy and sell these allowances on the carbon market, introducing flexibility and cost efficiency into the system. Firms that reduce their emissions below their allocated allowances can sell the surplus, while those that exceed their cap must purchase additional credits, thereby internalising the environmental cost of emissions and creating a financial incentive for decarbonisation (European Commission 2016).

The core principle behind cap-and-trade systems is to establish a financial incentive to reduce emissions where it is most cost effective. As highlighted by Fankhauser and Hepburn (2010), this mechanism enables companies to compare the marginal cost of reducing emissions within their operations against the market price of carbon allowances. If it is cheaper for a company to reduce emissions (e.g. by upgrading to cleaner technology or optimising energy use) than to purchase additional allowances, the company is incentivised to make those reductions. Conversely, if mitigation is expensive internally, the company can comply by buying allowances from others who can abate more cheaply. The flexibility of these mechanisms can lower compliance costs and encourage innovation, though their effectiveness depends heavily on policy design, enforcement, and transparency. Over-allocation of emission allowances, for instance, can suppress carbon prices and weaken the incentive to cut emissions (Ellerman et al. 2010). Similarly, price volatility, driven by economic cycles or policy uncertainty, can deter long-term investment in low-carbon technologies (Newell et al. 2013). Another concern is carbon leakage, where firms shift operations to countries with laxer emissions regulations, resulting in no net decrease in global emissions. Finally, cap-and-trade systems also face governance and enforcement issues, particularly in ensuring accurate monitoring, reporting, and verification (MRV) of emissions data. Without robust regulatory oversight, markets are vulnerable to speculative trading or manipulation, which may distort price signals and undermine public trust.

Like cap-and-trade systems, carbon-offset schemes are market-based tools aimed at addressing climate change. These schemes allow entities to compensate for their emissions by purchasing carbon credits generated by emission-reduction or removal projects elsewhere, such as reforestation or methane-capture initiatives (Hyams and Fawcett 2013). These offsets are often used in voluntary carbon markets, though some compliance systems permit limited offset use. Unlike cap-and-trade, offset schemes do not impose a direct cap on emissions but instead rely on high-quality projects to deliver climate benefits indirectly. The key distinction lies in the fact that cap-and-trade restricts total emissions at the source, whereas offsets enable emission balancing through external mitigation efforts (Corbera et al. 2009).

Carbon-finance derivative products, such as carbon futures, options, and swaps, have emerged to hedge against the volatility in the price of carbon allowances. These instruments allow market participants, including speculators and compliance buyers, to manage price risk and ensure cost predictability. For example, a company might lock in the future price of allowances through a futures contract to mitigate compliance-cost uncertainty. The expansion of carbon derivatives enhances market liquidity but also introduces complexities related to financial regulation and market oversight (Spilker and Nugent 2022).

1.2.7 Green Crowdfunding

Crowdfunding refers to an Internet-based mechanism for raising capital, where a large number of individuals (the "crowd") contribute small amounts of money through dedicated online platforms to support a project, venture, or cause (Cicchiello 2024). These platforms, known as crowdfunding platforms, serve as two-sided markets that match fundraisers and funders, facilitating not only financial transactions but also social validation and marketing (Belleflamme et al. 2015). Crowdfunding has recently emerged has a disruptive and participatory instrument within the expanding landscape of green finance (Dinh et al. 2024). Unlike traditional funding sources, such as banks, private equity, and venture capital funds, which often privileges large-scale, institutionally vetted projects, crowdfunding enables decentralised financing of green entrepreneurial initiatives by directly engaging individuals, both as investors and as advocates for environmental change (Corsini et al. 2024). This model democratises access to capital, particularly for small businesses and early-stage ventures pursuing social and environmental goals, which often face systemic barriers within traditional finance due to their higher intrinsic difficulty in producing tangible and measurable results in the short term (Petruzzelli et al. 2019). Crowdfunding encompasses various models, including investment-based (e.g. equity, debt, or royalty-based returns), reward-based (i.e. non-financial incentives such as early access or personalised products), and donation-based (i.e. philanthropic contributions with no tangible return). Within the green finance sector, investment-based crowdfunding, and the equity-based model in particular, is the most developed and impactful. In this case, investors, motivated by both financial returns and environmental impact, acquire small ownership stakes in green startups and small- to medium-sized enterprises (SMEs) in sectors such as renewable energy, circular economy, sustainable agriculture, and clean technology (Cicchiello et al. 2023b). Lending-based crowdfunding models, in which individual lenders provide loans directly to borrowers, are also active in green finance but less mature, often limited by trust, platform regulation, and borrower screening challenges (Jung and Lee 2022). The distinctive feature of green crowdfunding lies not only in its ability to bridge the financing gap that frequently constrains sustainable ventures, but also its capacity to foster collective agency and bottom-up innovation (Carè et al. 2025). Based on a participatory logic, crowdfunding platforms allow investors to become active participants in driving the transition towards a sustainable economy. In this way, green crowdfunding promotes a cultural shift in the way sustainability is funded and perceived (Butticè et al. 2019).

1.2.8 Blue Bonds and Blue Loans

Blue finance refers to a recently emerged subset of green finance that aims to mobilise public and private capital to fund activities that preserve and sustainably manage ocean, coastal, and freshwater ecosystems (Wang et al. 2024). Specifically, blue bonds and blue loans are innovative financing instruments focused on projects that contribute to the protection of oceans and sustainable water management (Choudhary et al. 2021). These financial instruments intent to contribute to the achievement of Sustainable Development Goals (SDG) 6 (clean water and sanitation: ensure availability and sustainable management of water and sanitation for all) and 14 (life below water: conserve and sustainably use the oceans, seas, and marine resources for sustainable development), respectively (International Finance Corporation 2022). Like green bonds, blue bonds are fixed-income instruments issued by governments, development banks, or others whose proceeds are used exclusively for financing or refinancing environmentally friendly projects focused on the blue economy (Thompson 2022). Similar to green bonds, issuing blue bonds often entails higher costs compared to conventional debt instruments. These additional expenses stem from the need to manage and monitor the use of proceeds, conduct detailed environmental impact assessments, and obtain external reviews or certifications to ensure transparency and accountability Shan et al. (2023).

The first blue bond ever issued was launched by the Republic of Seychelles in October 2018, marking a historic milestone in sustainable finance and ocean conservation Benzaken et al. (2024). The bond was partially guaranteed by the World Bank and included a concessional loan from the Global Environment Facility (GEF). The bond raised US$15 million from international investors with a 10-year maturity and a coupon of 6.5%. It was the world's first sovereign blue bond, specifically designed to support the sustainable management and expansion of Seychelles' marine protected areas, improve governance of priority fisheries, and develop the country's blue economy.

In 2021, Belize entered a groundbreaking $364 million blue bond transaction, structured as a debt-for-nature swap, representing one of the most ambitious applications of blue finance to sovereign debt management Standing (2023). Under this innovative arrangement, Belize repurchased a significant portion of its external debt, at a discount, using proceeds from a blue bond issued by Credit Suisse and backed by a political risk-insurance guarantee from the U.S. International Development Finance Corporation (DFC). In exchange, Belize has committed more than US$180 million to a marine conservation fund over the next 20 years. This funding is dedicated to protecting 30% of Belize's marine territory and coastal environments, including some of the most biodiverse coral reefs in the world (Fontana-Raina and Grund 2024).

Blue loans are specialised lending instruments, typically structured as bilateral or syndicated loans (non-tradable), making them accessible to both corporates and governments for medium-to-large-scale blue infrastructure or conservation investments (Thompson 2022). As a form of private financing (borrower-lender agreement), blue loans offer considerable flexibility in their structuring, making them adaptable to the diverse needs of borrowers across sectors such as coastal tourism, marine conservation, and ocean-based renewable energy (Narula et al. 2023). When tied to measurable environmental performance indicators, blue loans can also be structured as sustainability-linked instruments, adjusting repayment terms based on the borrower's progress towards predefined marine or climate-related targets. Blue bonds and blue loans are aligned with the Green Bond Principles (GBP) and Green Loan Principles (GLP), respectively, with the additional requirement that 100% of proceeds be allocated to clearly defined blue activities (International Finance Corporation 2022).

1.3 Greenwashing and Other Risks in Green Finance

As green finance markets grow rapidly, they present both vital opportunities and considerable risks that must be addressed to safeguard credibility, investor trust, and environmental integrity (Buchetti et al. 2024). One of the most pervasive risks is greenwashing, the practice of marketing financial products, corporate initiatives, or investments as environmentally friendly when they do not deliver measurable or meaningful sustainability outcomes. Greenwashing may involve vague or misleading product labels, selective disclosure, and reliance on unverified or inconsistent ESG data. Critically, intensifying market competition in the fast-growing sustainable finance sector is a key driver of greenwashing: As financial institutions and companies face mounting pressure to attract green capital and meet investor expectations, they may be tempted to overstate their environmental credentials (Cicchiello et al. 2023). In highly competitive markets, firms risk engaging in symbolic rather than substantive sustainability actions, seeking to differentiate themselves without achieving genuine impact (Delmas and Burbano 2011).

A complementary risk channel emerges from programme design. Evidence from the COVID-era public loan guarantees shows that, while guarantees facilitated a shift of credit towards low-emission firms and the formation of new bank-firm ties, banks' internal monitoring (e.g. PD updates) was relaxed and less predictive under guarantees—especially for low-emission borrowers—highlighting the higher information-production costs of green lending and the possibility that risk controls weaken when downside risk is shifted to the public sector (Buchetti et al. 2025). This underscores why integrity safeguards (eligibility criteria, ex post MRV, and independent assurance) must travel with any scaling instrument, whether labels or guarantees.

Closely tied to the problem of greenwashing is the lack of standardisation in green finance. Globally, different jurisdictions employ varying definitions of what constitutes "green" or "sustainable", and taxonomies are still evolving. The absence of universally accepted standards hampers investors' ability to compare products and assess risk consistently. The EU Taxonomy represents a significant step forward, but cross-border alignment remains a work in progress (Dumrose et al. 2022). Meanwhile, ESG scoring methodologies, often used to guide sustainable investment decisions, suffer from opacity and inconsistency. Major ESG rating agencies frequently employ divergent frameworks and data sources, resulting in significant variation in ESG scores for the same entity (Bissoondoyal-Bheenick et al. 2024). This inconsistency can mislead investors and result in "ESG arbitrage", where companies selectively highlight favourable metrics to secure lower financing costs.

Another systemic risk is market fragmentation, mainly due by the parallel development of market practices based on commercial priorities that produce divergent outcomes (Pyka 2023). The proliferation of green financial products, including green bonds, sustainability-linked instruments, and ESG funds, has created a complex and fragmented landscape. This complicates due diligence and can expose investors to unexpected risks, particularly where legal frameworks do not clearly define fiduciary duties related to sustainability claims. Reputational risk is also significant in green finance. Financial institutions that misrepresent their green credentials face growing scrutiny from regulators, civil society, and the public (Forcadell and Aracil 2017). The reputational fallout from allegations of greenwashing can result in legal challenges, regulatory penalties, and loss of investor confidence. In addition, regulatory risk is rising, as jurisdictions like the European Union and the U.S. Securities and Exchange Commission move towards stricter disclosure and anti-greenwashing standards. Finally, there are data and verification risks. Many sustainability claims rely on self-reported data from companies, which may be incomplete or biased (Kotsantonis and Serafeim 2019). Even where third-party verification is used, the quality and consistency of assurance providers vary widely, increasing the potential for false confidence in reported outcomes. Addressing these risks requires robust monitoring, reporting, and verification (MRV) frameworks and harmonised global standards. Without such safeguards, the promise of green finance could be compromised, delaying the transition to a truly sustainable economy.

References

A. Afzal, E. Rasoulinezhad, Z. Malik, Green finance and sustainable development in Europe. Econ. Research-Ekonomska Istraživanja **35**(1), 5150–5163 (2022)

I. Akomea-Frimpong, A.S. Kukah, X. Jin, R. Osei-Kyei, F. Pariafsai, Green finance for green buildings: a systematic review and conceptual foundation. J. Clean. Prod. **356**, 131869 (2022)

A. Amighini, P. Giudici, J. Ruet, Green finance: an empirical analysis of the green climate fund portfolio structure. J. Clean. Prod. **350**, 131383 (2022)

M. Andersson, P. Bolton, F. Samama, Hedging climate risk. Financ. Anal. J. **72**(3), 13–32 (2016)

M. Babic, Green finance in the global energy transition: actors, instruments, and politics. Energy Res. Soc. Sci. **111**, 103482 (2024)

P. Belleflamme, N. Omrani, M. Peitz, The economics of crowdfunding platforms. Inform. Econ. Policy **33**, 11–28 (2015)

D. Benzaken, J.P. Adam, J. Virdin, M. Voyer, From concept to practice: financing sustainable blue economy in lessons learnt from the Seychelles experience. Marine Policy **163**, 106072 (2024)

E. Bissoondoyal-Bheenick, S. Bennett, R. Lehnherr, A. Zhong, ESG rating disagreement: implications and aggregation approaches. Int. Rev. Econ. Financ. **96**, 103532 (2024)

L. Broadway, Should trustees invest in sustainability-linked bonds? Trusts Trustees **28**(4), 275–280 (2022)

B. Buchetti, L.X. Lin, I. Miquel-Flores, S. Perdichizzi, A. Reghezza, Loan guarantee and portfolio greening: evidence from European credit registers. ECB Working Paper Series 2916, European Central Bank, Frankfurt am Main (2025). Accessed Jan 2025

B. Buchetti, I. Miquel-Flores, S. Perdichizzi, A. Reghezza, The green-transition risk in Europe & the role of public guarantees. SUERF Policy Brief No. 904 (2024). Retrieved May 28, 2025.

V. Butticè, M.G. Colombo, E. Fumagalli, C. Orsenigo, Green oriented crowdfunding campaigns: their characteristics and diffusion in different institutional settings. Technol. Forecast. Soc. Change **141**, 85–97 (2019)

R. Calel, Carbon markets: a historical overview. Wiley Interdiscip. Rev. Climate Change **4**(2), 107–119 (2013)

G. Cardillo, A. Duqi, S. Perdichizzi, G. Torluccio, Sustainable and socially responsible finance. Finance Res Lett. **67**, 105878 (2024)

R. Carè, R. Fatima, M. Cerciello, S. Taddeo, Exploring the landscape of green crowdfunding: trends, themes, and insights from a bibliometric review. Finance Res. Lett. **77**, 107143 (2025)

P. Choudhary, M. Khade, S. Savant, A. Musale, M.S. Chelliah, S. Dasgupta, Empowering blue economy: from underrated ecosystem to sustainable industry. J. Environ. Manag. **291**, 112697 (2021)

A.F. Cicchiello, *Crowdfunding European Business*, 1st edn. (Routledge, London, 2024)

A.F. Cicchiello, M. Cotugno, C. Foroni, Does competition affect ESG controversies? Evidence from the banking industry. Financ. Res. Lett. **55**, 103972 (2023)

A.F. Cicchiello, M. Cotugno, S. Monferrà, S. Perdichizzi, Credit spreads in the European green bond market: a daily analysis of the covid-19 pandemic impact. J. Int. Financ. Manag. Account. **33**(3), 383–411 (2022a)

A.F. Cicchiello, M. Cotugno, S. Monferrà, S. Perdichizzi, Which are the factors influencing green bonds issuance? Evidence from the European bonds market. Financ. Res. Lett. **50**, 103190 (2022b)

A.F. Cicchiello, A. Gatto, D. Salerno, At the nexus of circular economy, equity crowdfunding and renewable energy sources: are enterprises from green countries more performant? J. Cleaner Prod. **410**, 136932 (2023b)

A.F. Cicchiello, F. Marrazza, S. Perdichizzi, Non-financial disclosure regulation and environmental, social, and governance (ESG) performance: the case of EU and US firms. Corporate Soc. Responsibility Environ. Manag. **30**(3), 1121–1128 (2023c)

Climate Bonds Initiative, Global state of the market 2024 (2025). Retrieved June 11, 2025.

E. Corbera, M. Estrada, K. Brown, How do regulated and voluntary carbon-offset schemes compare? J. Integrative Environ. Sci. **6**(1), 25–50 (2009)
F. Corsini, F.P. Appio, M. Frey, Green crowdfunding: an empirical study of success factors. IEEE Trans. Eng. Manag. **71**, 7654–7668 (2024)
M.A. Delmas, V.C. Burbano, The drivers of greenwashing. California Manag. Rev. **54**(1), 64–87 (2011)
C. Di Tommaso, S. Perdichizzi, S. Vigne, A. Zaghini, Is the government always greener? CFS Working Paper 718, Center for Financial Studies (2024)
D.W. Diamond, Financial intermediation and delegated monitoring. Rev. Econ. Stud. **51**(3), 393–414 (1984)
J.M. Dinh, A.J. Isaak, M.C. Wehner, Sustainability-oriented crowdfunding: an integrative literature review. J. Cleaner Prod. **448**, 141579 (2024)
K. Du, J. Harford, D.D. Shin, Who benefits from sustainability-linked loans? Finance Working Paper 917, European Corporate Governance Institute (2023)
M. Dumrose, S. Rink, J. Eckert, Disaggregating confusion? The EU taxonomy and its relation to ESG rating. Financ. Res. Lett. **48**, 102928 (2022)
T. Ehlers, D. Gao, F. Packer, A taxonomy of sustainable finance taxonomies. BIS Papers 118, Bank for International Settlements (2021)
A.D. Ellerman, F.J. Convery, C. De Perthuis, *Pricing Carbon: The European Union Emissions Trading Scheme* (Cambridge University Press, Cambridge, 2010)
European Commission, The EU emissions trading system (EU ETS) (2016). Retrieved June 5, 2025
S. Fankhauser, C. Hepburn, Designing carbon markets. Part I: carbon markets in time. Energy Policy **38**(8), 4363–4370 (2010)
P. Feldhütter, K. Halskov, A. Krebbers, Pricing of sustainability-linked bonds. J. Financ. Econ. **162**, 103944 (2024)
C. Flammer, Green bonds: effectiveness and implications for public policy. Environ. Energy Policy Econ. **1**(1), 95–128 (2020)
S. Fontana-Raina, S. Grund, Debt-for-nature swaps: the Belize 2021 deal and the future of green sovereign finance. Capital Markets Law J. **19**(2), 128–151 (2024)
F.J. Forcadell, E. Aracil, European banks' reputation for corporate social responsibility. Corp. Soc. Responsib. Environ. Manag. **24**(1), 1–14 (2017)
L.G. Giraudet, A. Petronevich, L. Faucheux, Differentiated green loans. Energy Policy **149**, 111861 (2021)
J. Giráldez, S. Fontana, Sustainability-linked bonds: the next frontier in sovereign financing. Capital Markets Law J. **17**(1), 8–19 (2022)
B. Holmström, J. Tirole, Financial intermediation, loanable funds, and the real sector. Q. J. Econ. **112**(3), 663–691 (1997)
K. Hyams, T. Fawcett, The ethics of carbon offsetting. Wiley Interdiscip. Rev. Climate Change **4**(2), 91–98 (2013)
G.I. Iacobuţă, C. Brandi, A. Dzebo, S.D.E. Duron, Aligning climate and sustainable development finance through an SDG lens: the role of development assistance in implementing the Paris agreement. Global Environ. Change **74**, 102509 (2022)
ICMA, Sustainability-linked bond principles (2024). Accessed 28 May 2025
International Finance Corporation, Guidelines for blue finance: guidance for financing the blue economy, building on the green bond principles and the green loan principles (2022). Accessed 27 May 2025
Y. Jung, K. Lee, Roles and tasks of peer-to-peer lending platforms in activating green finance, in *Green Digital Finance and Sustainable Development Goals*, ed. by F. Taghizadeh-Hesary, S. Hyun (Springer, Singapore, 2022), pp. 353–366
J.F. Kölbel, A.P. Lambillon, Who pays for sustainability? An analysis of sustainability-linked bonds (2022). Research Paper 23, Swiss Finance Institute
S. Kotsantonis, G. Serafeim, Four things no one will tell you about ESG data. J. Appl. Corp. Financ. **31**(2), 50–58 (2019)

M. Migliorelli, P. Dessertine, *The Rise of Green Finance in Europe. Opportunities and Challenges for Issuers, Investors and Marketplaces* (Palgrave Macmillan, Cham, 2019)
S.P. Mishra, P. Kumar, J. Rout, An overview of green, social, sustainability, and sustainability-linked (GSSS) bonds. J. Soc. Econ. Dev. **25**, 127–145 (2023)
A. Monk, R. Perkins, What explains the emergence and diffusion of green bonds? Energy Policy **145**, 111641 (2020)
K. Narula, M.J. Spalding, T. Thiele, J. Dyer, A. Pouponneau, Generating finance for blue economy transition (2023). Accessed 27 May 2025
R.G. Newell, W.A. Pizer, D. Raimi, Carbon markets 15 years after Kyoto: lessons learned, new challenges. J. Econ. Perspect. **27**(1), 123–146 (2013)
L.H. Pedersen, S. Fitzgibbons, L. Pomorski, Responsible investing: the ESG-efficient Frontier. J. Financ. Econ. **142**(2), 572–597 (2021)
S. Perdichizzi, B. Buchetti, A.F. Cicchiello, L. Dal Maso, Carbon emission and firms' value: evidence from Europe. Energy Econ. **131**, 107324 (2024)
A.M. Petruzzelli, A. Natalicchio, U. Panniello, P. Roma, Understanding the crowdfunding phenomenon and its implications for sustainability. Technol. Forecast. Soc. Change **141**, 138–148 (2019)
M. Pyka, The EU green bond standard: a plausible response to the deficiencies of the EU green bond market? Eur. Bus. Organ. Law Rev. **24**(4), 623–643 (2023)
S. Shan, N. Mirza, M. Umar, A. Hasnaoui, The nexus of sustainable development, blue financing, digitalization, and financial intermediation. Technol. Forecast. Soc. Change **195**, 122772 (2023)
G. Spilker, N. Nugent, Voluntary carbon market derivatives: growth, innovation & usage. Borsa Istanbul Rev. **22**, S109–S118 (2022)
A. Standing, The financialization of marine conservation: the case of debt-for-ocean swaps. Development **66**(1), 46–57 (2023)
J.E. Stiglitz, A. Weiss, Credit rationing in markets with imperfect information. Am. Econ. Rev. **71**(3), 393–410 (1981)
B.S. Thompson, Blue bonds for marine conservation and a sustainable ocean economy: status, trends, and insights from green bonds. Marine Policy **144**, 105219 (2022)
A. Trotta, Environmental impact bonds: review, challenges, and perspectives. Curr. Opin. Environ. Sustain. **66**, 101396 (2024)
S. Utz, Corporate scandals and the reliability of ESG assessments: evidence from an international sample. Rev. Manag. Sci. **13**, 483–511 (2019)
G. Vulturius, A. Maltais, K. Forsbacka, Sustainability-linked bonds–their potential to promote issuers' transition to net-zero emissions and future research directions. J. Sustain. Financ. Invest. **14**(1), 116–127 (2024)
D. Wang, X. Zhang, S. Chen, Visualizing the landscape of blue finance for sustainable development: a bibliometric analysis and future directions. Bus. Strategy Dev. **7**(2), e368 (2024)

Chapter 2
The Role of Green Bonds in the Fight Against Climate Change

Abstract This chapter traces how labelled green bonds moved from niche to systemic relevance, with issuance surging from USD 40 billion (2014) to over USD 550 billion (2023) and cumulative outstandings above USD 1.6 trillion. We document two pricing premia. First, a conditional "greenium" emerges in specific regimes and concentrates in ring-fenced, "dark-green" use of proceeds, with magnitudes shaped by liquidity, benchmark eligibility, disclosure credibility, and policy milestones (Paris Agreement, EU taxonomy acts, central-bank eligibility). Second, a Government Green Spread (public over private) widens counter-cyclically, reflecting benchmark-curve formation and flight-to-quality dynamics. Macro-financial frictions—exchange-rate volatility, capital-account constraints, and collateral policy—amplify or damp both premia, while local supply effects compress spreads until scarcity dissipates. A tranche-level study of 1389 European green bonds (2014–2021) shows rating as the dominant driver, an independent coupon-structure premium for floating-rate notes, and only limited size effects once sector is controlled; speculative-grade issuers do not enjoy a "green discount". On regulation, we contrast the EU's taxonomy-anchored EuGB framework and ESMA oversight with the U.S. reliance on principles and private certification, outlining gaps in standardisation, verification quality, and market readiness. The chapter closes with a research agenda on time-varying premia, primary-secondary linkages, cross-asset interactions, policy risk, digital market infrastructure, and measuring real-world additionality.

Keywords Green bonds · Greenium · Government Green Spread · Liquidity and benchmark eligibility · Floating-rate notes · EU Green Bond Standard · EU Taxonomy · Monetary policy and collateral · Macro-financial frictions · Primary-market pricing

2.1 Green Bond Market

The labelled green bond segment has evolved from a philanthropic experiment into a strategically relevant source of long-term capital: Primary issuance climbed from

A. F. Cicchiello, S. Perdichizzi, *The Role of Green Bonds in Finance*, SpringerBriefs in Finance, https://doi.org/10.1007/978-3-032-14840-7_2

barely USD 40 billion in 2014 to more than USD 550 billion in 2023, bringing the cumulative stock above USD 1.6 trillion and raising the segment's share of global bond outstandings to just over three per cent (CBI 2024). Two discrete accelerations explain this trajectory. The first followed the 2015 Paris Agreement and the European Union's initial Sustainable Finance Action Plan, when annual supply multiplied six-fold to USD 257 billion by 2019 and the issuer base broadened beyond multilateral development banks (MDBs) and Scandinavian utilities to include banks, real-estate companies, and investment-grade corporates; during the same window the euro overtook the US dollar as the reference funding currency, accounting for 42 per cent of global labelled flow versus 30 per cent for dollars and 9 per cent for onshore renminbi (Moro and Zaghini 2025). The second acceleration, in 2020–2021, coincided with pandemic stimulus, the European Green Deal, and the incorporation of environmental eligibility into the ECB's Asset Purchase Programme, the Federal Reserve's Secondary Market Corporate Credit Facility, and the Bank of England's Corporate Bond Purchase Scheme; yearly issuance breached USD 500 billion, the euro's lead widened, and public issuers—sovereigns, agencies, and sub-nationals—captured more than one-third of net new money (Di Tommaso et al. 2025). Europe now accounts for roughly USD 590 billion of outstanding labelled debt (4.7% of its domestic bond market), the United States for USD 398 billion (0.8%), and China for USD 141 billion; nonetheless, the labelled share of total issuance is markedly higher in emerging economies (about seven per cent) than in advanced economies (about two per cent), reflecting the intersection of infrastructure needs and concessional climate finance. Sectoral rotation has been equally abrupt: MDBs and utilities, which contributed more than 80 per cent of early flow, now provide less than 15 per cent, having ceded primacy first to financials and, by 2021, to public entities whose benchmark curves compress liquidity premia for private issuers (Ehlers and Packer 2017).

Primary-market spreads exhibit two statistically distinct premia. The generic *greenium*—defined as the yield concession investors accept for environmental alignment—averages minus 28 basis points in a 203 000-bond matched sample balanced on tenor, rating, coupon type, currency, and issuance date (Moro and Zaghini 2025); quarterly difference-in-differences estimates show the premium was indistinguishable from zero before 2019, widened sharply across the three quarters following March 2020 and stabilised near minus 30 basis points despite the subsequent 350 bp rise in global risk-free rates. Comparable magnitudes emerge from twin-bond pairs that differ only by colour, with euro-area sovereigns pricing at a minus three to minus six-basis-point differential relative to their brown twins (Bolton et al. 2022), and from corporate event studies that bracket the full-sample greenium between minus five and minus 35 basis points once liquidity controls are imposed (Baker et al. 2022). Cross-sectional dispersion is left-skewed: Trading-frequency regressions attribute roughly two-thirds of variance to liquidity and benchmark-eligibility effects, while the fat left tail is concentrated in high-volatility currencies and subordinated or callable structures (Kapraun and Scheins 2019). Only bonds with ring-fenced proceeds for "dark-green" assets—renewables, certified net-zero buildings, zero-emission transport—earn significant premia; instruments with imprecise or mixed-use sustainability labels price close to conventional debt, confirming investors' aversion to impact-verification risk. High-frequency identification around ICMA guideline revisions, EU taxonomy delegated acts, and sovereign climate-bill votes corroborates the causality of the label: Abnormal spread compressions cluster within event windows narrower than one trading day, ruling out macro-confounds (Ehlers and Packer 2017).

Issuer identity generates an additional premium. The *Government Green Spread* (GGS)—the difference between public-sector and private-sector green coupons—averages minus 48 basis points across 5 905 labelled bonds and deepened to almost minus 70 basis points in 2019–2021 (Di Tommaso et al. 2025). Quarterly interaction coefficients reveal a counter-cyclical pattern: The spread is negligible in tranquil periods, widens during spikes in the Caldara-Iacoviello geopolitical-risk index and the Gavriilidis U.S. climate-policy-uncertainty index, and narrows again once volatility subsides. Triple-interaction specifications show the GGS remains materially negative even for issuers in the highest tertile of debt-to-GDP or the lowest tertile of per-capita income, suggesting that balance-sheet credibility, benchmark-curve formation, and flight-to-quality dynamics dominate macro-fundamental weakness. Three mechanisms rationalise the hierarchy. First, sovereigns internalise transition externalities and can cross-subsidise long-duration mitigation projects, lowering perceived renegotiation risk (OECD 2023). Second, public benchmarks reduce search and execution costs for investors, compressing the liquidity premia embedded in corporate curves. Third, macro-shock episodes trigger allocations into safe assets, and green-labelled sovereigns satisfy that demand while preserving portfolio-level climate targets. Eligibility decisions amplify these channels: A triple-difference study around the ECB's Pandemic Emergency Purchase Programme shows purchase-eligible green bonds tightening an additional 22 basis points relative to ineligible labelled peers and 35 basis points relative to brown comparators, with the effect persisting for at least six months and coinciding with statistically significant improvements in issuer carbon disclosure (Zaghini 2024).

Macro-financial frictions modulate both premia. A dynamic two-country model matched to 30000 observations links greenium size inversely to exchange-rate volatility and directly to capital-account frictions: A five-percentage-point rise in the standard deviation of monthly FX returns halves the observed discount by lifting conventional yields, whereas a five-percentage-point inflow tax deepens the discount by up to 15 basis points conditional on investor risk aversion (Moro and Zaghini 2025). Free-float regimes erode approximately half the greenium and nearly all of the GGS observed under pegs, explaining why labelled issuance is proportionally larger—but dearer in absolute terms—in Brazil, South Africa, and Turkey (Moro and Zaghini 2025). Supply effects matter too: Panel estimates indicate that a one-percentage-point increase in the quarterly share of green issuance compresses average spreads by two to four basis points, but premia plateau once the local labelled share approaches ten per cent of outstanding debt, signalling diminishing marginal scarcity (Caramichael and Rapp 2024). At firm level, announcement-return studies document positive abnormal equity gains of one to two per cent, though the magnitude attenuates in markets with weak disclosure mandates or contested environmental policy, consistent with higher verification costs (Tang and Zhang 2020).

Four stylised facts therefore emerge. *First*, labelled supply is strongly policy elastic: Each milestone intervention—Paris, EU taxonomy, central-bank eligibility—coincides with discrete jumps in both volume and market share. *Second*, investors pay a persistent premium for environmental alignment once credit quality, tenor, and liquidity are controlled; the greenium is robustly centred around minus 20-to-minus 35 basis points across identification strategies. *Third*, public issuers command an additional, counter-cyclical concession of 40 to 70 basis points, validating their catalytic role in deepening markets and anchoring price discovery. *Fourth*, exchange-rate regimes, capital-account openness, and collateral policy can amplify or damp both premia, giving fiscal, monetary, and prudential authorities tangible levers for steering the cost of capital towards net-zero objectives. The coexistence of robust investor demand, benchmark-induced liquidity benefits, and demonstrable policy sensitivity indicates that green bonds are now large enough to influence portfolio allocation, price efficient enough to transmit environmental information and responsive enough to be guided by well-designed regulation.

2.2 Regulatory Challenges and Levers

In recent years, green bonds have emerged as a crucial instrument for financing environmentally responsible projects and, consequently, for supporting the successful transition to a low-carbon economy (Sartzetakis 2021). The rapid proliferation of green bonds has drawn attention to the need for robust regulatory frameworks to foster market growth and investor confidence, ensuring the credibility and comparability of green bonds (Piñeiro-Chousa et al. 2021).

One of the main challenges to the development of the green bond market has been the lack of a commonly accepted definition of green bonds, which has raised investor concerns about "greenwashing" practices (de Freitas Netto et al. 2020). Without a standardised global definition, issuers may label bonds as green based on inconsistent or self-serving criteria, undermining the credibility of the market. Furthermore, transparency remains uneven, particularly in post-issuance reporting and the use of proceeds, which limits stakeholders' ability to evaluate environmental impact. In many markets, third-party verification is voluntary, and disclosure practices vary widely. This regulatory gap underscores the need for enforceable accountability mechanisms to ensure that issuers adhere to environmental commitments over time (Flammer 2021).

The European Union has taken significant steps to establish a coherent regulatory framework for sustainable finance, particularly with regard to green bonds. The initiative began with the *Action Plan on Financing Sustainable Growth* launched by the European Commission in 2018, which identified the need to reorient capital flows towards sustainable activities, manage financial risks stemming from climate change, and foster transparency and long termism in financial and economic activity (European Commission 2018).

A major milestone was achieved with the adoption of Regulation (EU) 2020/852, commonly referred to as the EU Taxonomy Regulation, which came into force in July 2020. This regulation established a unified classification system to determine whether an economic activity qualifies as environmentally sustainable. It defined six environmental objectives: climate change mitigation, climate change adaptation, sustainable use and protection of water and marine resources, transition to a circular economy, pollution prevention and control, and protection and restoration of biodiversity and ecosystems (European Union 2020). The taxonomy provides the necessary definitional backbone for all subsequent EU-level green finance instruments, including green bonds.

Building upon this foundation, the European Commission proposed the regulation on the European Green Bond Standard (EuGBS) in July 2021. This proposal aimed to create a voluntary but robust framework for labelling bonds as "European Green Bonds", provided that their proceeds were fully aligned with the EU Taxonomy (European Commission 2021). The EuGBS includes stringent pre- and post-issuance transparency requirements and mandates external reviews by accredited verifiers.

The proposal was formalised in September 2023 with the adoption of Regulation (EU) 2023/2631 by the European Parliament and Council, which officially established the European Green Bond Standard. This regulation introduces the "EuGB" label and sets out detailed rules on use-of-proceeds disclosures, sustainability reporting, and verification processes. To ensure quality assurance and accountability, it also empowers the European Securities and Markets Authority (ESMA) to oversee and supervise external reviewers (European Parliament and Council 2023).

ESMA's *Final Report on Technical Standards on the European Green Bonds Regulation* was published on 14 February 2025, laying the foundation for the operationalisation of the EuGB regulatory framework (European Securities and Markets Authority 2025). This final report has been developed based on the feedback and recommendations received from the Securities and Markets Stakeholder Group (SMSG) and the Proportionality and Coordination Committee (PCC) during the public consultation that closed in June 2024.

Together, these initiatives reflect the EU's holistic and phased approach to regulating green bonds. By combining voluntary standards with rigorous alignment to the EU Taxonomy and strong supervisory mechanisms, the framework aims to foster a credible green finance ecosystem that can serve as a model globally.

Table 2.1 provides a chronological overview of the key regulatory developments shaping the European green bond market. Starting with the European Commission's 2018 *Action Plan on Financing Sustainable Growth*, the table outlines each milestone, the year of adoption, and the institution responsible.

Despite regulatory progress, the EU still faces several challenges, including the voluntary nature of the EU GBS and the complexity of aligning national frameworks with the general taxonomy. Regulatory fragmentation remains a concern, as individual member states often implement sustainability disclosure requirements

Table 2.1 European green bond regulatory timeline

Year	Regulatory milestone	Issuing entity
2018	Action Plan on Financing Sustainable Growth	European Commission
2020	EU Taxonomy Regulation (Reg. EU 2020/852)	European Parliament and Council
2021	Proposal for EU Green Bond Standard (EuGBS)	European Commission
2023	Adoption of Reg. (EU) 2023/2631 on EuGBS	European Parliament and Council
2024–2025	Delegated Acts and Technical Standards	European Commission, ESMA

Source: Authors' elaboration

inconsistently (Pyka 2023; Gargantini 2024). Moreover, a gap persists between regulatory ambition and market readiness, particularly for smaller issuers who may lack the resources to comply with extensive reporting obligations (Christensen et al. 2021).

In contrast, the United States lacks a centralised green bond regulation (Flammer 2020). Instead, the market operates primarily through self-regulation and third-party certifications such as those provided by the Climate Bonds Initiative (CBI) or adherence to the ICMA Green Bond Principles (Park 2018). While this flexibility has facilitated market growth, it also introduces inconsistencies in how green projects are defined and verified (Wang 2018). The absence of federal regulatory oversight limits standardisation, and discussions around a U.S. green taxonomy or federal disclosure requirements remain in early stages (Ballester et al. 2024). Among the most significant levers for regulatory advancement in both regions are the development of unified taxonomies, mandatory impact reporting, and enhanced due-diligence requirements. Additionally, international coordination, especially through forums such as the International Platform on Sustainable Finance (International Platform on Sustainable Finance 2023), can help mitigate fragmentation and promote global alignment. Effective regulation must balance stringency with practicality, ensuring that green bonds remain both credible and accessible. In this context, the evolving regulatory landscape will play a pivotal role in shaping the future of sustainable finance, both in Europe and globally.

2.3 Key Factors Affecting Green Bond Issuance: A Review of the Literature

The issuance of green bonds has evolved into a dynamic financial mechanism through which firms and public entities fund environmentally sustainable initiatives. The motivation behind this trend extends beyond traditional financing needs, encompassing strategic, reputational, and institutional drivers (Löffler et al. 2021). The growing volume of scholarly attention reflects this complexity, revealing that green bond issuance is often used to signal environmental commitment, meet stakeholder expectations, and leverage emerging investor preferences for sustainability (e.g. Bhutta et al. 2022; Kedia and Joshipura 2023).

Central to the theoretical understanding of green bond issuance is *signalling theory* (Connelly et al. 2011; Hartman-Glaser 2017). Firms that issue green bonds differentiate themselves by voluntarily adhering to higher environmental disclosure standards, thereby reducing information asymmetry between managers and investors (Zhu et al. 2024). The credibility of such signals is typically reinforced by certification mechanisms, such as third-party verification or alignment with the Green Bond Principles, which validate the environmental purpose of the bond's proceeds (Lyon and Maxwell 2011; Lyon and Montgomery 2015). As Flammer (2021) argues, green bond issuance is not only a financial tool but also a strategic communication mechanism aimed at stakeholders, including investors, regulators, and civil society.

Empirical studies offer robust evidence supporting this signalling role. Gianfrate and Peri (2019), for instance, find that firms benefit reputationally from green bond issuance, translating into improved investor perception and even lower cost of capital, consistent with the existence of a "greenium". This yield discount, which Zerbib (2019) quantifies in the range of two–six basis points on average, reflects investors' willingness to sacrifice return in favour of environmental impact. These effects are not uniform across markets and issuers; rather, they depend on contextual factors such as investor base, regulatory stringency, and the credibility of the issuer's environmental claims.

Regulatory frameworks and national policy environments significantly influence issuance dynamics. Tolliver et al. (2020) highlight how supportive policies, such as tax incentives, standardised taxonomies, and green investment mandates, enhance green bond market development by lowering entry barriers and aligning issuer incentives. These findings are echoed by Lin and Su (2022), who demonstrate that high environmental performance, governance quality, and corporate social responsibility scores are predictive of green bond issuance, especially in countries with mature financial systems.

Moreover, corporate governance plays a pivotal role. Firms with diverse and independent boards tend to issue green bonds more frequently, reflecting an ethical orientation and stronger risk oversight. Board diversity, particularly female representation, has been associated with a higher likelihood of green financial activity, suggesting that inclusiveness in decision-making correlates with sustainability-oriented corporate behaviour (Cicchiello et al. 2022).

Issuance motivations also vary by sector and firm characteristics (Fatica and Panzica 2021; Flammer 2021). Larger firms, especially in capital-intensive industries like energy and infrastructure, are more inclined to issue green bonds due to their scale, visibility, and access to long-term capital markets. These issuers often align green bonds with their broader sustainability strategies, using the instrument to fund projects that reflect their corporate environmental goals. Lin and Su (2022) demonstrate that firms with formal sustainability committees and robust ESG frameworks are significantly more likely to issue green bonds, suggesting a strategic alignment rather than opportunistic behaviour. Additionally, Tolliver et al. (2020)

find that capital-intensive firms, especially those in energy and infrastructure, adopt green bonds as tools for financing long-term environmental investment, further evidencing their integration into corporate sustainability planning.

The earmarking of proceeds is another critical feature that enhances green bond credibility. Earmarked funds, allocated exclusively to green projects, are central to the trust-building mechanism that reassures investors about environmental impact.

Market conditions and investor behaviour additionally shape issuance patterns. While early-stage green bonds faced liquidity constraints, more recent studies (e.g. Zerbib 2019) indicate that increased market depth, standardisation, and issue size have enhanced liquidity, reducing one of the initial barriers to issuance. Furthermore, the evolving profile of green bond investors, typically institutions with long-term horizons and ESG mandates, has created a reliable source of demand, which in turn influences firm-level decisions about green bond issuance.

Importantly, peer pressure and reputational benchmarking within industries also drive adoption. Firms may respond to competitor behaviour and market expectations by issuing green bonds, especially when operating in sectors under intense environmental scrutiny or public attention (Bedendo et al. 2023; Kim et al. 2025). This dynamic creates a positive feedback loop: The more prevalent green bond issuance becomes, the more it is perceived as a norm or necessity in sustainable finance.

In conclusion, the literature underscores that green bond issuance is not a monolithic phenomenon but a multifaceted strategic choice. It is informed by firm-specific traits, investor preferences, governance structures, institutional support, and the broader regulatory environment. The signalling of environmental commitment, backed by credible disclosure and external verification, remains a central pillar in understanding why firms opt for green finance over conventional instruments.

2.3.1 An Empirical Study of the European Market

2.3.1.1 Data, Variable Construction, and Descriptive Statistics

A comprehensive tranche-level dataset of *1389* euro-denominated green bonds issued between the first quarter of 2014 and the fourth quarter of 2021 by entities domiciled in the European Economic Area and the United Kingdom is assembled. Primary-market observations originate from Bloomberg's New Issues monitor, while contractual attributes—coupon structure, maturity, face value, and syndicate composition—are matched from Refinitiv's Terms and Conditions service. To mitigate false positives, green status is cross-checked with the Climate Bonds Initiative (CBI) taxonomy; tranches that lack CBI verification or allude merely to "ESG" or "sustainability" without an explicit use-of-proceeds link to climate mitigation are excluded.[1]

[1] CBI verification, coupled with prospectus keyword searches, yields a concordance rate of 96.3 per cent. Remaining discrepancies—mostly private placements—amount to fewer than 2 per cent of raw observations and are dropped.

Table 2.2 Summary statistics, 2014–2021

Variable	*N*	Mean	SD	Min	p25	Median	p75	Max
Yield at issuance (%)	1389	0.932	0.721	–0.137	0.392	0.791	1.288	4.248
Maturity (years)	1389	9.842	6.263	1	5	8	12	100
Deal size (USD mill)	1389	87.4	192.4	5	25	50	100	1200

Yields are gross re-offering rates. Deal size is expressed in 2021 U.S. dollars using ECB reference rates; inflation effects are negligible

Five screening filters are subsequently applied. *First*, currency homogeneity is enforced: Only euro-denominated liabilities are retained, thereby sidestepping FX-induced level shifts in yield. *Second*, perpetuals, convertibles, asset-backed structures, and notes with embedded call options whose first-call date precedes five years are removed to avoid pricing distortions linked to optionality. *Third*, duplicate CUSIP–ISIN pairs are reconciled across data providers to eliminate double counting. *Fourth*, retail "mini-bonds" of face value below 5 million, frequently placed off index and devoid of pricing transparency, are eliminated. *Fifth*, the dependent variable—gross re-offering yield—is expressed in percentage points and *trimmed at the 5th and 95th percentiles*.

The resulting balanced panel retains 94 per cent of the raw Bloomberg universe and covers 28 domiciles, 10 primary sectors, and six rating buckets. Descriptive moments are reported in Table 2.2. The mean coupon (0.93 per cent) lies approximately one standard deviation below the mean of euro investment-grade corporate issuance over the same horizon, underscoring the sustained effect of European Central Bank (ECB) asset purchase programmes. At the same time, the standard deviation of 72-basis-point signals economically meaningful heterogeneity, which provides statistical power for the cross-sectional regressions that follow.

Figure 2.1 plots the arithmetic mean coupon by calendar year. A monotonic decline from roughly 2.0 per cent in 2014 to 0.8 per cent in 2020 is evident, followed by a mild rebound to approximately 1.0 per cent in 2021. The trajectory coincides with the ECB's Asset Purchase Programme (APP), its Corporate Sector Purchase Programme (CSPP), and the subsequent Pandemic Emergency Purchase Programme (PEPP) taper; Altavilla et al. (2021) document a comparable 120-basis-point compression in swap-adjusted investment-grade spreads over the same horizon. The granularity of the decline suggests that green bonds, as a sub-asset class, remained largely integrated with the broader euro fixed-income market rather than carving out a segregated "greenium" channel.

Figure 2.2 depicts the trimmed density. The mode lies at 0.4 per cent, and 62 per cent of tranches price below 1 per cent. Nonetheless, a right tail extending to 4 per cent remains, confirming the presence of speculative-grade issuers even in the highly regulated labelled segment. Kurtosis of 11.7, relative to a Gaussian benchmark of 3, indicates heavy-tailed behaviour. Importantly, the persistence of the right tail after trimming attests that the heterogeneity is structural rather than a mere artefact of data noise.

Fig. 2.1 Average yield at issuance by year

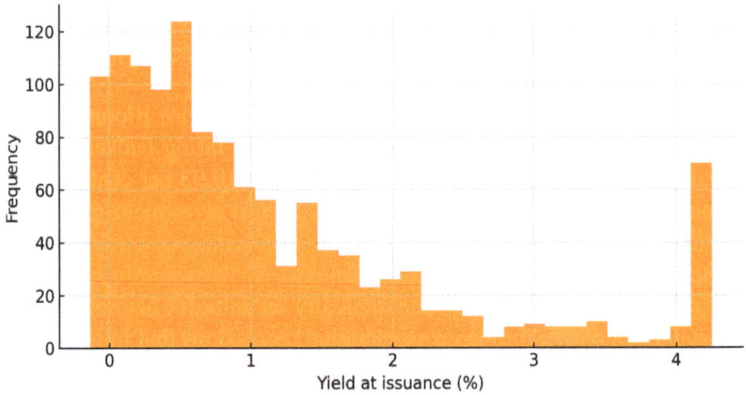

Fig. 2.2 Distribution of primary-market yields

2.3.1.2 Issuer Breadth: Geography and Sector

The diversity of issuer domicile is summarised in Table 2.3 (*Issuance Counts by Country*). Germany and France alone account for approximately one-third of the sample. Nonetheless, peripheral sovereign exposure remains significant, with Spain, Italy, and Portugal contributing a combined 14 per cent. This breadth of participation contradicts the common conjecture that green borrowing is primarily concentrated among core AAA-rated sovereigns and their affiliated agencies.

Parallel insights emerge on the sectoral dimension. Table 2.4 (*Issuance Counts by Sector*) highlights a strong concentration in capital-intensive sectors: Utility and energy firms alone account for 238 tranches. However, green issuance is far from sectorally narrow. Bank-related issuers—municipal, provincial, and commercial—contribute 484 tranches in total, while government-sponsored enterprises and

Table 2.3 Issuance counts by country

Country	Count	Country	Count	Country	Count	Country	Count
Germany	383	Finland	38	Estonia	10	Romania	2
France	307	Denmark	33	Lithuania	10	Malta	1
Spain	153	Portugal	29	Croatia	7	Bulgaria	1
Netherlands	99	Norway	24	Iceland	6	Turkey	1
Italy	66	Greece	17	Slovakia	6		
Luxembourg	64	Ireland	15	Latvia	5		
United Kingdom	56	Poland	12	Cyprus	4		
Belgium	55	Czech Republic	11	Hungary	4		
Austria	43	Sweden	45	Slovenia	2		

Table 2.4 Issuance counts by sector

Panel A: top 17 sectors		Panel B: remaining sectors	
Sector	Count	Sector	Count
Bank Prov. and Munic.	266	Health care	14
Utility and Energy	238	Consumer non-cyc.	11
Commercial banks	218	Technology	9
Dev./multilaterals	101	Media and entertainment	8
Govt-sponsored Ent.	83	Insurance	8
Transportation	76	Metals and mining	7
Real estate (REIT)	64	Chemicals	6
Local authorities	60	Hotels and Leisure	6
Real Estate (Property)	51	Paper & Forest Prod.	5
Construction	47	Aerospace and Defense	4
Industrial Mfg.	38	Utilities (water)	4
Telecom	27	Retail	3
Consumer cyclical	25	Textiles and Apparel	3
Agriculture/food	19	Diversified industrials	3
		Auto manufacturing	2
		Packaging	2
		Pharmaceuticals and biotech	2
		Oil and Gas	2
		Shipbuilding	1
		Other	1

transportation entities together add another 159. In fact, nearly half of all deals originate from these broader public and quasi-public institutional spheres.

The distribution thus affirms that the green bond label has expanded well beyond early adopters in the utilities sector. Participation now spans a diverse range of industries, from real estate and manufacturing to telecommunications and agriculture, reflecting the mainstreaming of green finance across both geography and economic activity.

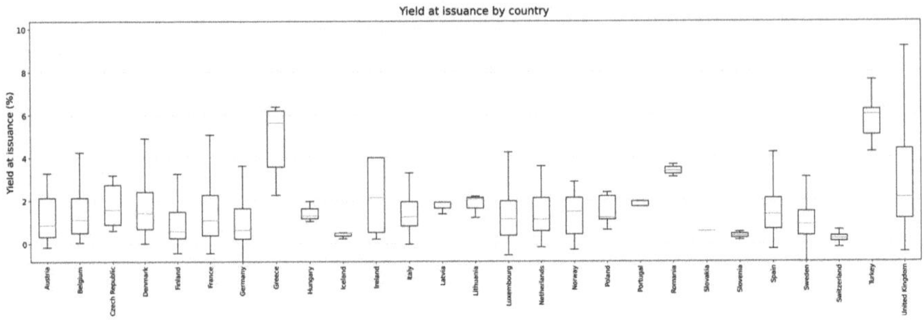

Fig. 2.3 Yield at issuance by issuer country

Figure 2.3 illustrates the cross-country distribution of gross re-offering yields for euro-denominated *green bonds* issued between 2014 and 2021. Median yields remain below *0.60 per cent* in core economies such as Finland and Germany, increase to approximately *1.15 per cent* in Luxembourg, and range between *0.90* and *1.15 per cent* in Spain and Italy—indicating a persistent, though somewhat reduced, core-periphery spread within the monetary union.

Interquartile ranges are relatively narrow for AAA/AA-rated sovereigns (≤ 0.8 percentage points), but widen markedly for peripheral issuers: Greece and the Czech Republic exhibit spreads exceeding *1.5 percentage points*, while Turkey's upper whisker extends beyond *5.5 per cent*, even after winsorisation.

This distribution suggests that, within the labelled segment, green bonds largely mirror traditional sovereign-risk stratification. Yield differentials continue to reflect underlying fiscal fundamentals, covenant robustness, and the maturity of domestic investor bases—rather than a uniform "greenium".

Sectoral price dispersion is depicted in Fig. 2.4. Multilateral development banks (MDBs) and government-sponsored entities price at or below the euro swap curve, whereas transportation companies and real-estate investment trusts (REITs) face medians above 2 per cent. The ordering is congruent with the risk spectrum posited by Bolton and Kacperczyk (2023), where brown-capex sectors bear higher spreads due to transition uncertainty.

2.3.1.3 Maturity, Liquidity, and Coupon Modality

Figure 2.5 plots yield against time to maturity. The slope is nearly flat—about eight basis points per additional year—suggesting that the term premium is markedly compressed in the euro area.

Figure 2.6 shows a weak inverse relation between yield and log-deal size. In univariate space, larger tranches enjoy a three-basis-point discount relative to micro issues; nonetheless, multivariate regressions indicate that the coefficient loses significance once sector fixed effects are included, implying that the liquidity advantage of size is overwhelmingly sector-bound (Table 2.5).

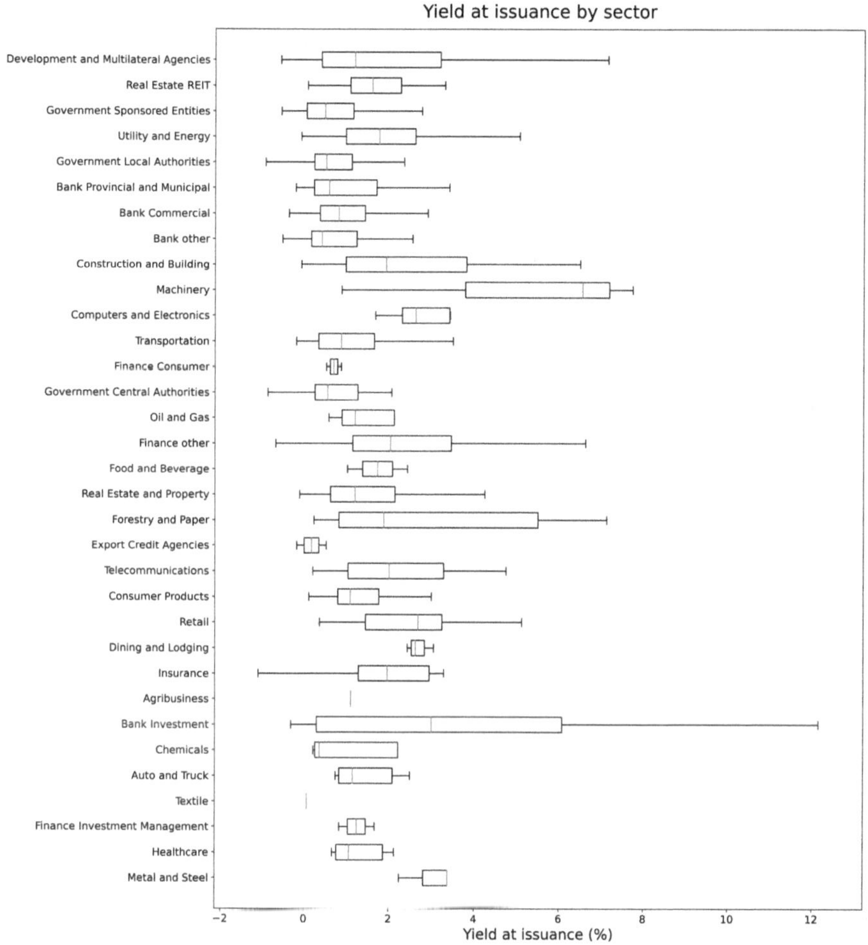

Fig. 2.4 Yield at issuance by sector

Table 2.6 indicates that only 70 tranches (5 per cent) employ floating coupons. Figure 2.7 reveals a median premium of approximately 60 basis points for floating-rate notes (FRNs) relative to fixed-rate bonds. The premium persists after conditioning on rating, maturity, and sector, thereby implying an autonomous coupon-structure effect.

2.3.1.4 Econometric Specification and Results

Primary-market pricing determinants are examined via the cross-sectional OLS model

$$Y_i = \alpha + \beta_1 \text{Maturity}_i + \beta_2 \log(\text{Size}_i) + \sum_j \gamma_j \mathbf{D}_{ij}^{\text{Rating}} + \delta\, \text{FRN}_i + \sum_k \theta_k \mathbf{D}_{ik}^{\text{Sector}} + \varepsilon_i,$$

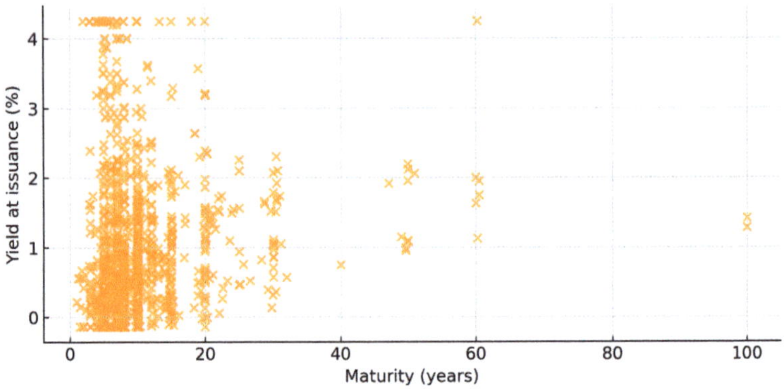

Fig. 2.5 Yield and time to maturity

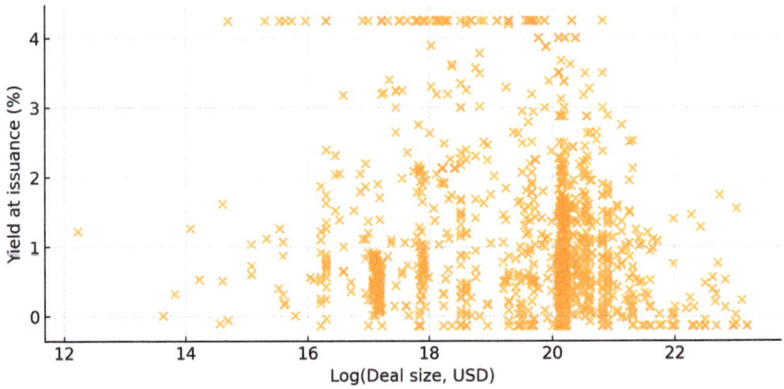

Fig. 2.6 Yield and deal size

Table 2.5 Yield regressions: bivariate slopes and full specification

	Dependent variable: yield at issuance (%)			
	(1)	(2)	(3)	(4)
Maturity (years)	0.078*** (0.021)	–	0.072*** (0.023)	0.071** (0.032)
log(Deal size)	–	–0.030** (0.014)	–0.011 (0.016)	–0.006 (0.015)
Floating-rate dummy	–	–	–	0.421***
Rating dummies	–	–	–	Yes
Sector dummies	–	–	–	Yes
R^2	0.033	0.018	0.041	0.292
Obs.	1 389	1 389	1 389	1 230

Notes: Robust standard errors clustered by issuer country in parentheses. Significance levels: ***, **, * denote 1%, 5%, and 10%, respectively

Table 2.6 Coupon-type composition

Coupon type	Count
Fixed rate	1319
Floating rate	70

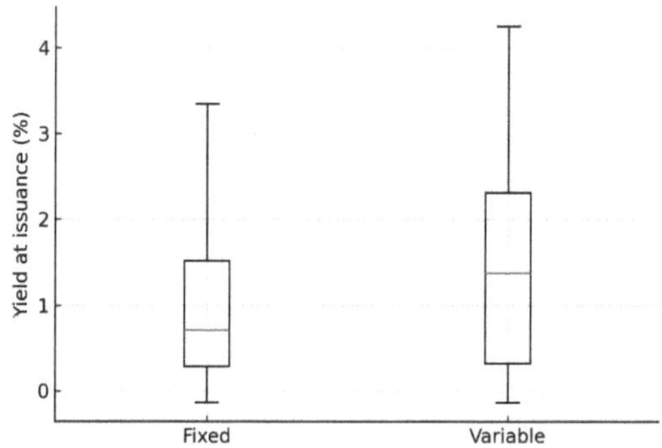

Fig. 2.7 Yield by coupon type

Table 2.7 Determinants of primary-market yield

Variable	Coef.	Std. Err.	t	p
Intercept	2.558	0.647	3.95	0.000
AA	−0.231	0.133	−1.74	0.083
AAA	−0.006	0.210	−0.03	0.977
BB	1.414	0.189	7.48	0.000
BBB	0.050	0.158	0.31	0.754
Maturity (years)	0.071	0.032	2.22	0.027
log(deal size)	−0.006	0.015	−0.42	0.674
Floating-rate dummy	0.421	0.148	2.85	0.005
Sector fixed effects	Yes			
R^2	0.29			

where Y_i denotes the trimmed gross re-offering yield of bond i; $\mathbf{D}^{\text{Rating}}$ comprises five dummies with single-A as the baseline; and $\mathbf{D}^{\text{Sector}}$ encompasses nine sector dummies with Utility and Energy as the omitted group. Standard errors are clustered at the issuer-country level to absorb correlated placement practices and local demand shocks (Petersen 2009). Results are summarised in Table 2.7.

Three principal insights emerge. *First*, the BB dummy is economically dominant, adding 141 basis points relative to single-A, thereby confirming that speculative-grade issuers do not enjoy a "green discount". *Second*, the maturity coefficient remains positive and significant despite the aggregate flatness of the curve, indicating that residual duration risk is priced once credit and sector are controlled for. *Third*, the floating-rate dummy retains a premium of 42 basis points, corroborating the coupon-structure effect visualised earlier. The magnitude aligns with Fleckenstein and Longstaff (2020), who document a 35- to 50-basis-point FRN premium in the U.S. investment-grade market even after controlling for duration.

A growing body of issuance data reveals a clear two-track pricing regime. High-grade, fixed-rate borrowers—chiefly sovereigns, supranationals, agencies, and investment-grade utilities—price within statistical noise of the euro swap curve, whereas speculative-grade names and floating-rate notes routinely command sizeable yield premia. The long-lived 40- to 60-basis-point surcharge attached to FRNs appears driven by either (i) the thin buy-and-hold demand for variable-rate cash flows among ESG-constrained investors or (ii) heightened model risk surrounding reference-rate resets in an extended low-for-long environment.

This coupon-structure effect carries immediate regulatory and market-design consequences. First, the forthcoming EU Green Bond Standard (EU-GBS), by enforcing harmonised reporting and verification across coupon formats, has the potential to compress the current differential. Second, the European Banking Authority's draft proposal to grant preferential regulatory capital treatment to green senior-non-preferred debt is likely to narrow speculative-grade spreads, particularly for the BB cohort that presently absorbs the heaviest credit penalty within the labelled segment. Third, the findings caution against taking headline yield compression as prima facie evidence of a universal "greenium": Neglecting to account for rating bucket, sector affiliation, and coupon archetype materially overstates the price advantage attributed to the green label.

Collectively, these results show that primary-market pricing of euro-denominated green bonds reflects the confluence of macro-level policy forces and bond-specific microstructure. Negative policy rates and sustained quantitative-easing anchor the overall level of yields, yet credit quality, industry grouping, and coupon architecture jointly explain the remaining cross-section. Although credit rating remains the single strongest driver, coupon form commands an independent premium that the prevailing green bond discourse too often ignores. Rigorous evaluation of a "greenium" therefore must control for conventional bond covariates, while the disclosure-standardisation initiatives now progressing through EU rulemaking appear poised to accelerate convergence between fixed- and floating-rate structures.

2.4 Research on Green Bonds: Reflections and Future Directions

The preceding review and empirical evidence point to a market that has outgrown its initial niche and now interacts endogenously with monetary policy, regulatory architecture, and macro-financial conditions. Against this backdrop, the next wave of scholarship should broaden both its empirical lenses and its impact metrics. We identify six research avenues that follow organically from the stylised facts documented in this chapter:

1. **Time-varying pricing dynamics.** The chapter shows that the average *greenium* is neither uniform nor permanent. Future work should employ rolling-window or state-space models to capture how premia respond to supply surges, quantitative-easing cycles, and exogenous commodity-price shocks. High-frequency identification around EuGBS milestones and ECB collateral announcements remains largely unexplored.

2. **Primary-secondary-market linkages.** While most studies, including ours, focus on launch-day spreads, the determinants of *secondary-market* liquidity, bid-ask volatility, and transaction costs are still poorly understood. Order-book data and TRACE-style post-trade tapes could reveal whether label-induced liquidity benefits persist once bonds season or whether they dissipate as information asymmetry subsides.
3. **Cross-asset and portfolio interactions.** Empirical tests of green bonds as diversifiers rarely control for the evolving correlation structure between labelled debt, carbon-intensive equities, and transition-hedging instruments (e.g. EU ETS futures). Factor-model and copula approaches can clarify how green bonds perform under extreme-value stress and whether they improve downside-risk metrics at the portfolio level.
4. **Regulatory uncertainty and policy risk.** Event-study and difference-in-discontinuities designs can quantify how regulatory fragmentation, delegated-act delays, or sudden policy reversals re-price labelled debt relative to brown comparators.
5. **Technological innovation and market infrastructure.** Blockchain-native issuance platforms promise immutable use-of-proceeds tracking and automated coupon step-ups for covenant breaches. Micro-case studies and pilot-programme RCTs could test whether such features compress information-verification costs enough to widen the investor base or deepen liquidity.
6. **Real-world impact beyond financial metrics.** Finally, the field must move from *pricing* green finance to *assessing its environmental and social additionality*. Linking bond-level proceeds to granular emissions data, biodiversity indicators, or measures of climate-justice distribution would require interdisciplinary collaboration and novel data-fusion techniques, but it is indispensable for evaluating whether green bonds accelerate the transition rather than merely cheapen it.

Taken together, these directions call for empirical designs that are multi-dimensional, policy aware, and impact-oriented—bridging the disciplines of finance, sustainability science, and governance studies. Progress along these fronts will determine whether green bonds remain a complementary funding tool or evolve into a systemic conduit for climate-aligned capital allocation.

References

C. Altavilla, G. Carboni, R. Motto, Asset purchase programs and financial markets: lessons from the euro area. Int. J. Central Banking **17**(5), 1–48 (2021)

M. Baker, D. Bergstresser, G. Serafeim, J. Wurgler, The pricing and ownership of U.S. green bonds. Annu. Rev. Financ. Econ. **14**(1), 415–437 (2022)

L. Ballester, A. González-Urteaga, L. Shen, Green bond issuance and credit risk: international evidence. J. Int. Financ. Markets Instit. Money **94**, 102013 (2024)

M. Bedendo, G. Nocera, L. Siming, Greening the financial sector: evidence from bank green bonds. J. Bus. Ethics **188**(2), 259–279 (2023)

U.S. Bhutta, A. Tariq, M. Farrukh, A. Raza, M.K. Iqbal, Green bonds for sustainable development: review of literature on development and impact of green bonds. Technol. Forecast. Soc. Change **175**, 121378 (2022)

P. Bolton, L. Buchheit, M. Gulati, U. Panizza, B. Weder di Mauro, J. Zettelmeyer, Climate and debt. Geneva Reports on the World Economy, 25 (2022)

P. Bolton, M. Kacperczyk, Global pricing of carbon-transition risk. J. Financ. **78**(6), 3677–3754 (2023)

J. Caramichael, A.C. Rapp, The green corporate bond issuance premium. J. Bank. Financ. **162**, 107126 (2024)

CBI, Sustainable debt: global state of the market 2024 (2024)

H.B. Christensen, L. Hail, C. Leuz, Mandatory CSR and sustainability reporting: economic analysis and literature review. Rev. Account. Stud. **26**(3), 1176–1248 (2021)

A.F. Cicchiello, M. Cotugno, S. Monferrà, S. Perdichizzi, Which are the factors influencing green bonds issuance? Evidence from the European bonds market. Financ. Res. Lett. **50**, 103190 (2022)

B.L. Connelly, S.T. Certo, R.D. Ireland, C.R. Reutzel, Signaling theory: a review and assessment. J. Manag. **37**(1), 39–67 (2011)

S.V. de Freitas Netto, M.F.F. Sobral, A.R.B. Ribeiro, G.R.D.L. Soares, Concepts and forms of greenwashing: a systematic review. Environ. Sci. Europe **32**(1), 19 (2020)

C. Di Tommaso, S. Perdichizzi, S. Vigne, A. Zaghini, Is the government always greener? Working-paper version on EconStor (2025)

T. Ehlers, F. Packer, The growing market for green bonds: finance and certification. BIS Q. Rev. (2017)

European Commission, Action plan: financing sustainable growth (2018). Accessed 22 July 2025

European Commission, Proposal for a regulation on European green bonds (2021). Accessed 22 July 2025

European Parliament and Council, Regulation (EU) 2023/2631 on European green bonds and optional disclosures for bonds marketed as environmentally sustainable and for sustainability-linked bonds (2023). Accessed 24 July 2025

European Securities and Markets Authority, Final report: technical standards on the European green bonds regulation (2025). ESMA84-858037815-195 Accessed 24 July 2025

European Union, Regulation (EU) 2020/852 on the establishment of a framework to facilitate sustainable investment (2020). Accessed 22 July 2025

S. Fatica, R. Panzica, Green bonds as a tool against climate change? Bus. Strategy Environ. **30**(5), 2688–2701 (2021)

C. Flammer, Green bonds: effectiveness and implications for public policy. Environ. Energy Policy Econ. **1**(1), 95–128 (2020)

C. Flammer, Corporate green bonds. J. Financ. Econ. **142**(2), 499–516 (2021)

M. Fleckenstein, F.A. Longstaff, The US treasury floating rate note puzzle: is there a premium for mark-to-market stability? J. Financ. Econ. **137**(3), 637–658 (2020)

M. Gargantini, Regulatory harmonization and fragmentation in the capital markets union, in *Balancing Unity and Diversity in EU Legislation* (Edward Elgar, 2024), pp. 135–154

G. Gianfrate, M. Peri, The green advantage: exploring the convenience of issuing green bonds. J. Cleaner Prod. **219**, 127–135 (2019)

B. Hartman-Glaser, Reputation and signaling in asset sales. J. Financ. Econ. **125**(2), 245–265 (2017)

International Platform on Sustainable Finance, Annual report 2023 (2023). Accessed 28 July 2025

J. Kapraun, C. Scheins, Incredibly green: which bonds trade at a green premium (2019). Working Paper Goethe University Frankfurt, Germany

N. Kedia, M. Joshipura, Green bonds for sustainability: current pathways and new avenues. Manag. Financ. **49**(6), 948–974 (2023)

S.S. Kim, V. Nugroho, L. Handoko, Competitions among sub-financial sectors and growth of green bond markets in ASEAN plus three countries. Manag. Financ. **51**(1), 166–184 (2025)

B. Lin, T. Su, Green bond vs conventional bond: outline the rationale behind issuance choices in China. Int. Rev. Financ. Anal. **81**, 102063 (2022)
T.P. Lyon, J.W. Maxwell, Greenwash: corporate environmental disclosure under threat of audit. J. Econ. Manag. Strategy **20**(1), 3–41 (2011)
T.P. Lyon, A.W. Montgomery, The means and end of greenwash. Organ. Environ. **28**(2), 223–249 (2015)
K.U. Löffler, A. Petreski, A. Stephan, Drivers of green bond issuance and new evidence on the "greenium". Eurasian Econ. Rev. **11**(1), 1–24 (2021)
A. Moro, A, Zaghini, The green sin: how exchange–rate volatility and financial openness affect green premia. Rev. Financ. **29**(4), 1189–1217 (2025)
OECD, Sovereign borrowing outlook 2023: focus on green and social bonds (2023)
S.K. Park, Investors as regulators: green bonds and the governance challenges of the sustainable finance revolution. Stanford J. Int. Law **54**, 1–46 (2018)
M.A. Petersen, Estimating standard errors in finance panel data sets: comparing approaches. Rev. Financ. Stud. **22**(1), 435–480 (2009)
J. Piñeiro-Chousa, M. López-Cabarcos, J. Caby, A. Šević, The influence of investor sentiment on the green bond market. Technol. Forecast. Soc. Change **162**, 120351 (2021)
M. Pyka, The eu green bond standard: a plausible response to the deficiencies of the EU green bond market? Eur. Bus. Organ. Law Rev. **24**(4), 623–643 (2023)
E.S. Sartzetakis, Green bonds as an instrument to finance low carbon transition. Econ. Change Restruct **54**(3), 755–779 (2021)
D.Y. Tang, Y. Zhang, Do shareholders benefit from green bonds? J. Financ. Econ. **135**(2), 635–653 (2020)
C. Tolliver, A.R. Keeley, S. Managi, Drivers of green bond market growth: the importance of nationally determined contributions to the Paris agreement and implications for sustainability. J. Cleaner Prod. **244**, 118643 (2020)
E.K. Wang, Financing green: reforming green bond regulation in the United States. Brooklyn J. Corp. Financ. Commercial Law **12**(2), 9–52 (2018)
A. Zaghini, Unconventional green. J. Corp. Financ. **85**, 102556 (2024)
O.D. Zerbib, The effect of pro-environmental preferences on bond prices: evidence from green bonds. J. Banking Financ. **98**, 39–60 (2019)
Q. Zhu, X. Zhao, M. Wu, Third-party certification: how to effectively prevent greenwash in green bond market? Analysis based on signalling game. Environ. Dev. Sustain. **26**(6), 16173–16199 (2024)

Chapter 3
Green Versus Brown Bonds: Are They Really Priced Differently?

Abstract This chapter compares green and conventional ("brown") bonds along purpose of proceeds, transparency and post-issuance reporting, verification, pricing, liquidity, risk, investor base, and market behaviour. We clarify why use-of-proceeds discipline and independent reviews differentiate labelled debt and synthesise mixed evidence on the "greenium": While investors sometimes accept lower yields for credible, ring-fenced issues, effect sizes vary by issuer type, sector, regime, and disclosure quality and often vanish when fundamentals are controlled. Using a large euro-denominated panel, we estimate issuance-day pricing with rating, sector, size, maturity, and time fixed effects and show that the average greenium is statistically indistinguishable from zero; observed unconditional gaps are largely explained by conventional covariates and supply-demand conditions. Liquidity advantages accrue to larger, benchmark-eligible issues; diversification benefits are limited given high correlation with conventional fixed income, although downside protection may appear in specific stresses. We discuss theoretical channels (non-pecuniary preferences, information asymmetry, perceived risk) and derive policy implications: Credible taxonomies, mandatory post-issuance impact reporting, and robust assurance are necessary for any persistent pricing differential; monetary and collateral frameworks can reinforce integrity by aligning incentives. Overall, pricing differences are conditional and earned by verification, not labels.

Keywords Greenium · Use of proceeds · Certification and verification · Liquidity and benchmark eligibility · Investor preferences and signalling · Fixed-effects regression · Euro-denominated bonds · EU Green Bond Standard · Primary-market pricing · Disclosure and impact reporting

3.1 Difference Between Green and Conventional Bonds

The rapid growth of green bonds as a financing tool for sustainable investments has led to extensive research on how they differ from conventional bonds across several areas such as the purpose of proceeds, transparency and reporting, pricing, risk, liquidity, investor base, and market dynamics (Hacıömeroğlu et al. 2022).

A. F. Cicchiello, S. Perdichizzi, *The Role of Green Bonds in Finance*, SpringerBriefs in Finance, https://doi.org/10.1007/978-3-032-14840-7_3

One of the most fundamental distinctions between green bonds and conventional bonds lies in the purpose of proceeds (Sangiorgi and Schopohl 2021). Green bonds explicitly commit the funds raised to financing or refinancing projects that deliver verifiable environmental benefits. This is formalised in international standards such as the Green Bond Principles (GBP) published by the International Capital Market Association (ICMA 2021) and the EU Green Bond Standard, which stipulate that proceeds must be exclusively allocated to eligible green projects (Tolliver et al. 2019). In contrast, conventional bonds typically impose no such restrictions, with proceeds used for general corporate or governmental purposes. Closely linked to the exclusive destination of the proceeds to green projects and activities is the dimension of transparency and reporting. Green bond issuers are required to disclose not only the intended use of proceeds, but also periodic reports on the allocation of funds and environmental impacts throughout the bond's life. The Green Bond Principles (ICMA 2021) and the Climate Bonds Initiative (CBI 2024) standards require issuers to publish post-issuance impact reports and periodic updates, which are essential to sustaining investor trust and mitigating concerns about "greenwashing". In contrast, conventional bonds lack such requirements, and investors generally receive no detailed information about the environmental use of the capital raised. Certification and verification provide another key distinction. While conventional bonds rely solely on issuer disclosures, many green bonds undergo independent external review processes, such as second-party opinions, certification under the CBI's Climate Bonds Standard, or other verifications, prior to issuance and during the bond's lifetime. Certification helps reduce information asymmetries and signals to the market that the bond aligns with accepted sustainability standards. According to CBI data (Climate Bonds Initiative 2024), over 86% of green bonds issued globally in recent years had some form of external review. These mechanisms of purpose restriction, transparency, and third-party verification fundamentally differentiate green bonds from conventional bonds and strengthen the credibility of the green bond market.

One prominent area of difference is pricing, specifically the existence of a "greenium" (i.e. a green bond premium). Empirical studies suggest that green bonds are often issued with lower yields compared to conventional counterparts, reflecting investors' willingness to accept reduced financial returns in exchange for supporting environmentally beneficial projects (Fatica et al. 2021; Caramichael and Rapp 2024). However, the magnitude and even existence of a greenium remain contested (MacAskill et al. 2021). Some studies argue that after controlling for bond characteristics (size, maturity, currency, rating), the greenium largely disappears or even reverses (Larcker and Watts 2020; Li et al. 2022). Given the importance of this topic for both theory and practice, the existence and determinants of the greenium will be examined in detail in the next paragraph.

Risk profile is another dimension where differences emerge. The long-term risk of green bonds is perceived to be lower due to alignment with sustainability goals and lower exposure to transition and regulatory risks (Sangiorgi and Schopohl 2021). However, some studies (Hacıömeroğlu et al. 2022) show that green bonds might exhibit higher maturity risk, as investors' confidence may decline if the use of proceeds is unclear over longer durations. Regarding liquidity, green bonds historically traded at lower liquidity levels due to market immaturity, smaller issue sizes, and limited secondary-market depth (Koziol et al. 2022). However, as the market matures, larger issue sizes and the entry of repeat issuers are enhancing green bond liquidity (Febi et al. 2018; Yan et al. 2025). For large and investment-grade green bonds, the liquidity gap with conventional bonds is now much narrower, although liquidity differentials persist for smaller or niche issues.

Differences also manifest in investor base and demand dynamics. Green bonds attract a distinct segment of retail and institutional investors motivated by ESG goals rather than purely financial returns Han and Li (2022). Survey evidence from European asset managers shows that strong green credentials and transparent reporting are decisive factors for green bond investment decisions Sangiorgi and Schopohl (2021). Poor reporting leads investors to avoid or divest such bonds. Moreover, green bonds are particularly attractive to long-term investors such as pension funds and insurance companies.

Beyond structural and pricing distinctions, green bonds differ from conventional bonds along several dimensions related to market behaviour. One of these dimensions is their correlation with broader financial markets. Empirical evidence indicates that green bonds are generally highly correlated with conventional corporate and Treasury bonds, especially for investment-grade issues (Tolliver et al. 2019). This limits their diversification benefits, although some segment-specific variation may exist for more niche or project-specific instruments. A second dimension is sensitivity to financial and geopolitical turbulence. During periods of market stress, green bonds may behave differently from conventional bonds depending on investor sentiment and ESG preferences. Some studies suggest that green bonds may offer partial downside protection during market shocks, as ESG-oriented investors are less prone to panic selling (Reboredo 2018; Naeem et al. 2021; Cicchiello et al. 2022a). However, risk spillovers from conventional markets to green bonds do increase during period of market stress (Pham 2016). The sensitivity of green bonds to geopolitical risks remains an area of emerging research interest, with systematic reviews highlighting the need for further investigation into how such risks may influence green bond pricing and investor behaviour (MacAskill et al. 2021; Koziol et al. 2022).

Geographical differentiation is also notable: European markets dominate global green bond issuance and pricing dynamics (Cicchiello et al. 2022b). The existence and magnitude of the greenium are more pronounced in Europe, where regulatory frameworks, investor demand, and green finance infrastructure are more developed (Fatica et al. 2021). In contrast, evidence from emerging markets suggests weaker greenium effects and more limited transparency or certification practices (Hu et al. 2022).

Finally, issuer motivations differ substantially between green and conventional bonds (Sangiorgi and Schopohl 2021; Dutordoir et al. 2024). Issuers of green bonds are often driven not only by financial considerations but also by reputational benefits, which can confer competitive advantage. Issuing a certified green bond can enhance a firm's environmental credibility and align with internal sustainability goals (Flammer 2021). In addition, by issuing green bonds companies can signal environmental innovation, attract ESG-focused investors, and access regulatory incentives. Moreover, governments and supranational institutions may issue green bonds as a tool to meet national or international climate commitments, rather than to optimise yield or minimise the cost of capital.

Table 3.1 summarises the key structural and market-related differences between green bonds and conventional bonds. Green bonds are distinguished by their environmental purpose, transparency requirements, and often by pricing and investor base characteristics. However, many of these differences are context-dependent and vary across markets, issuer types, and regulatory environments (Zerbib 2019; Wu 2022).

Table 3.1 Main differences between green and conventional bonds

Dimension	Green bonds	Conventional bonds
Purpose of proceeds	Finance projects with positive environmental impact (e.g. renewable energy)	No restrictions; general corporate, or government purposes
Transparency and reporting	High transparency required; must report on use of proceeds and environmental outcomes	No environmental reporting required
Certification/verification	Often third-party certified (e.g. Climate Bonds Standard, GBP alignment)	No certification related to ESG impact
Pricing/yield (Greenium)	Often lower yields (greenium) observed, but size and existence vary by market	Market-driven yield; no ESG-driven pricing effects
Liquidity	Historically lower liquidity; improving with market maturity and larger issue sizes	Generally higher liquidity, particularly for larger issuers
Investor base	Attracts ESG-focused investors, long-term institutional investors (pension funds, insurers)	Broad investor base, including short-term and yield-driven investors
Risk profile	Perceived lower transition and regulatory risks; may have higher maturity risk if project impacts unclear	Standard risk profile based on issuer creditworthiness
Correlation with markets	Correlated with conventional bond and energy markets; limited diversification benefit	Correlated with corporate and sovereign bond markets
Sensitivity to turbulence	May exhibit some downside protection during financial turbulence due to ESG investor preferences	Standard behaviour in line with credit and market risk
Geographical differentiation	Greenium and market depth stronger in Europe; weaker in emerging markets	No systematic geographical differentiation
Issuer motivations	Reputation signalling, regulatory incentives, ESG investor demand	General financing needs

Source: Authors' elaboration

3.2 Theoretical Background on the Existence (or Nonexistence) of a "greenium"

The existence of a "greenium", a yield discount on green bonds relative to comparable conventional bonds, has attracted substantial academic interest, given its implications for sustainable finance theory and the incentives driving green bond markets (MacAskill et al. 2021). Theoretically, a greenium may arise through two main channels: non-pecuniary motives and lower perceived risks. First, environmentally motivated investors may be willing to accept lower returns on green bonds, seeing them as tools to advance climate goals and to hedge against long-term transition risks (Zerbib 2019). Second, green bonds may be perceived as lower-risk assets, due to enhanced transparency, monitoring, and alignment with future regulatory landscapes (Löffler et al. 2021). Empirical evidence on the greenium is highly heterogeneous across markets, issuer types, and time periods (e.g. MacAskill et al. 2021). A large body of studies reports a positive greenium (see, for example, Hu et al. 2022; Koziol et al. 2022; Okafor et al. 2024). Löffler et al. (2021) document a greenium of 15–20 basis points on both primary and secondary markets, driven by investor preferences and information asymmetry reduction. Hu et al. (2024) find that in the Chinese green bonds market, both issuer greenness (i.e. the expected environmental benefits) and issue-level certification contribute to a measurable greenium. According to the authors, the issue's and issuer's perceived greenness mitigate information asymmetry between issuers and investors and thereby lower bond yields as investors are willing to pay a greenium for this greenness. In another study discussing the pricing of green bonds, Fatica et al. (2021) show that greenium magnitude depends on issuer type, with supranational and corporate green bonds benefiting from yield discounts, while green bonds from financial institutions do not. Similarly, Teti et al. (2022) document that green bonds in the Italian primary and secondary markets have an issuance spread that is 35 to 40 basis points lower than that of comparable conventional bonds, regardless of the issuer's sector. Recent studies further emphasise the role of market dynamics. Caramichael and Rapp (2024), for instance, find that global corporate greenium effects have strengthened since 2019 due to rising institutional investors' demand. Cicchiello et al. (2022a) report that during the COVID-19 crisis, European green bonds initially underperformed but later outperformed conventional bonds in response to positive news about the vaccines. Moreover, Huang et al. (2023) show that the greenium is non-linear, emerging more clearly in high-spread environments. Conversely, several papers provide strong counterevidence. Larcker and Watts (2020), using U.S. municipal bond data, report no systematic greenium, even among certified bonds. They argue that pro-environmental investor preferences are unlikely to explain prior findings and caution that certification alone does not produce a yield advantage. Wu (2022) likewise find no significant negative green bond premium in either Chinese or global markets when controlling for liquidity and certification effects. Finally, Li et al. (2022) show that only officially certified labelled green bonds in China lower financing costs of issuers, while the "green" characteristics alone are insufficient.

Further nuances emerge regarding liquidity effects and investor signalling. While some theories posit that greenium arises from enhanced liquidity, empirical studies frequently find no systematic liquidity differences between green and conventional bonds after controlling for issuer and bond characteristics (MacAskill et al. 2021). Instead, the signalling of credible environmental commitment, through third-party certification, transparent use of proceeds, and issuer reputation, has been identified as a more consistent driver of greenium formation (e.g. Löffler et al. 2021).

Finally, the literature increasingly stresses that greenium is a conditional phenomenon. It varies across issuer types, market segments, bond maturities, and macroeconomic contexts (MacAskill et al. 2021). Greenium effects are typically stronger in European sovereign and investment-grade segments (Cicchiello et al. 2022a; Koziol et al. 2022), while they are weaker or absent in emerging markets or lower-rated corporate debt (Wu 2022; Hu et al. 2022).

In summary, from what emerges from the literature, it is possible to state that, although a greenium exists under specific conditions, it is not universal. Variability across markets, issuer types, and bond characteristics persists. From a theoretical perspective, this suggests that future bond pricing models should incorporate non-financial investor preferences and governance quality (MacAskill et al. 2021). In practice, the greenium remains a potential but conditional incentive for issuers and investors, subject to a credible green signalling.

3.3 An Empirical Analysis of "greenium"

The rapid expansion of labelled debt markets has given rise to the widely held intuition that *green* bonds may offer issuers a measurable cost-of-capital advantage by attracting investors with climate-oriented preferences. Whether such an advantage continues to exist in a deeper, more transparent, and information-rich market environment remains an open empirical question—particularly when isolating the effect of the *bare label*, stripped of any additional signalling from detailed ESG disclosures.

This section addresses that question directly in the context of euro-denominated bond issuance. We pose three nested research questions:

1. Do European green bonds still command a yield discount at issuance—i.e. a *greenium*—even as annual supply now exceeds €400 billion?
2. To the extent that a greenium exists, is it uniform across maturities, issuance sizes, credit rating classes, sectors, or over time?
3. If the raw label no longer drives pricing differences, what policy tools or market design features could re-establish a cost-of-capital advantage for sustainable investments?

Our findings are as follows: (i) green bonds no longer enjoy a significant yield discount in primary markets; (ii) any marginal greenium observed in earlier years has fully dissipated by 2024; and (iii) only a robust institutional framework—combining stringent post-issuance disclosure, alignment with credible taxonomies, and collateral eligibility within monetary operations—offers a pathway to restoring a durable pricing benefit for green-labelled instruments.

Data Assembly and Coding Choices

Our dataset merges the BLOOMBERG New Issues Monitor with REFINITIV Terms and Conditions, capturing all euro-denominated tranches whose legal documentation was filed in the European Economic Area (plus the United Kingdom) and whose settlement dates fall between 1 January 2014 and 31 January 2024. Primary-market yields are sourced from Bloomberg's `Yield` field or Refinitiv's `Yield return` field and cross-checked against offering circulars when available.

We retain only fixed- or floating-rate instruments that meet the following criteria:

- Denominated in euro
- No exotic optionality (e.g. convertibles, reverse FRNs, step-ups)
- Face value of at least €5 million
- Stated maturity between 1 and 100 years
- Tranche-level pricing observable on launch day

A tranche is classified as green (`green = 1`) only if:

- 100% of net proceeds are earmarked for environmental purposes.
- The deal is listed by the Climate Bonds Initiative.

Manual keyword validation (*renewable, mitigation, taxonomy*) confirms a 96.3% match rate; residual mismatches are excluded. All other tranches are coded as conventional (`green = 0`).

Covariates included in the regression framework are:

- Maturity (in years)
- Log of deal size (in 2021 euros, deflated using Eurostat HICP)
- Best-available S&P or Moody's rating (five ordinal buckets, single-A as baseline)
- ICB Level-2 sector dummies (27 categories; Utilities as reference)
- Floating-rate dummy
- Calendar-quarter fixed effects

After applying all filters, the sample contains 2 326 tranches: 1 241 green and 1 085 conventional, totalling approximately €850 billion in issuance. Issuers span 28 sovereign jurisdictions and over 40 economic sectors. Descriptive statistics by label appear in Table 3.2.

Figure 3.1 overlays kernel-smoothed yield densities by label. The two curves nearly coincide, with green bonds showing a slightly thinner mode and comparable right-tail exposure, suggesting limited pricing differentiation in unconditional terms.

Figures 3.2 and 3.3 explore conditional heterogeneity. At the country level (Fig. 3.2), Nordic issuers (e.g. Sweden, Norway) tend to price green tranches *above* the local brown median, whereas Southern European markets (e.g. Spain, Italy) show a modest negative gap. Sectoral patterns (Fig. 3.3) are mixed: Utilities show negligible differences, but transportation exhibits a mild pricing advantage for green-labelled bonds.

Table 3.2 Descriptive statistics of primary-market yields

	Conventional	Green
Observations	2085	1241
Mean (%)	2.123	1.991
Std. dev. (%)	2.247	2.141
Min (%)	– 1.00	– 0.58
25th perc. (%)	0.48	0.53
Median (%)	1.40	1.25
75th perc. (%)	3.24	2.90
Max (%)	20.00	18.33

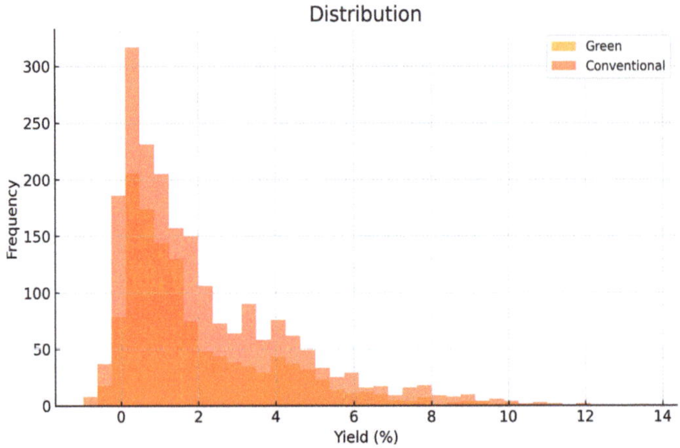

Fig. 3.1 Overlay of kernel-smoothed yield distributions by label

Temporal dynamics are visualised in Fig. 3.4, which plots monthly average yields by label and the resulting greenium (in basis points). A significant pricing gap emerges between 2017 and 2019—reaching up to –15 bp—but vanishes after 2021 as interest rates rise. This suggests that greenium effects, where present, are regime-dependent and potentially sensitive to macro-financial conditions.

Econometric Identification

To investigate whether bonds labelled as green enjoy a pricing advantage at issuance—a so-called "greenium"—we estimate the following cross-sectional regression:

$$\text{Yield}_i = \alpha + \beta\,\text{Green}_i + \gamma_1\,\text{Maturity}_i + \gamma_2\,\log(\text{Size})_i + \varepsilon_i, 1$$

where Yield_i denotes the gross yield of bond i at issuance (in basis points), Green_i is a dummy variable indicating whether the bond carries a green label, Maturity_i captures the bond's time to maturity in years, and $\log(\text{Size})_i$ denotes the logarithm of the tranche size. In this specification, the coefficient β estimates the average yield

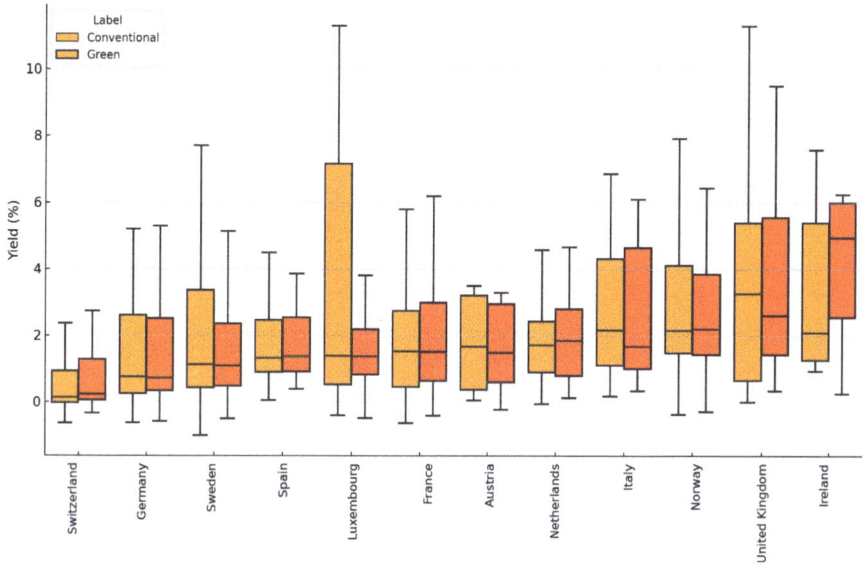

Fig. 3.2 Yield at issuance by issuer country (paired box plots)

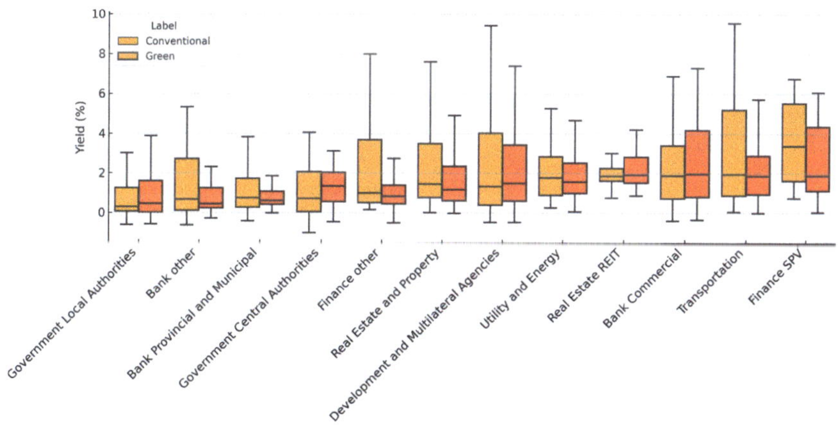

Fig. 3.3 Yield at issuance by sector (paired box plots)

differential—i.e. the greenium—between green and non-green bonds, controlling for key observable bond characteristics.

To account for additional sources of yield heterogeneity, we augment the baseline model with fixed effects for credit rating, economic sector, and calendar quarter in a second specification:

$$\text{Yield}_i = \alpha + \beta\,\text{Green}_i + \gamma_1\,\text{Maturity}_i + \gamma_2\,\log(\text{Size})_i + \delta_{\text{Rating}} + \delta_{\text{Sector}} + \delta_{\text{Quarter}} + \varepsilon_i.$$

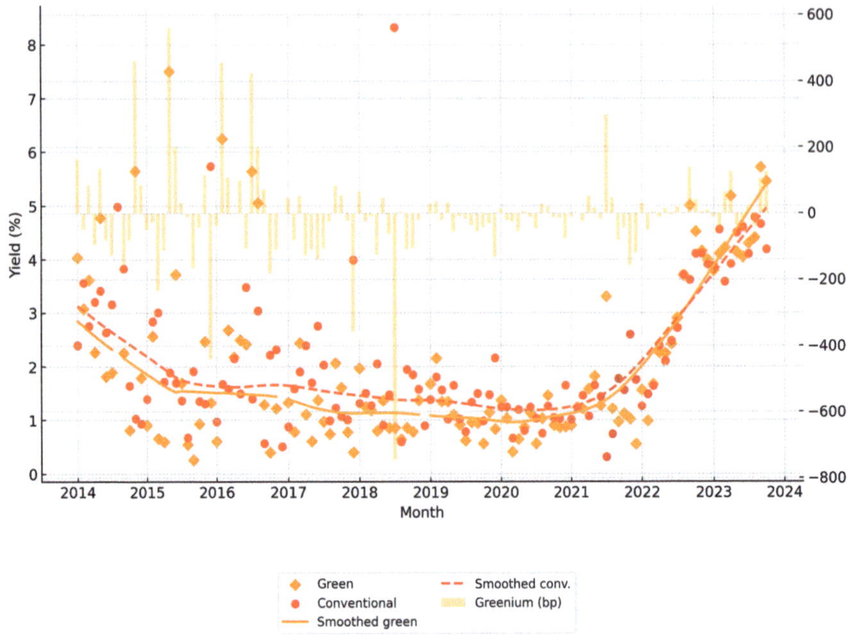

Fig. 3.4 Monthly average yields and greenium

Table 3.3 Green label and primary-market pricing

	(1) Baseline	(2) Full FE
Green dummy	− 0.146*	− 0.067
	(0.078)	(0.063)
Maturity (years)	− 0.002	0.023***
	(0.005)	(0.004)
log(size)	− 0.187***	− 0.276***
	(0.024)	(0.022)
Rating FEs	–	Yes
Sector FEs	–	Yes
Quarter FEs	–	Yes
R^2	0.020	0.456
Observations	3326	3326

Dependent variable: gross yield at issuance (%). Robust (Eicker-White) standard errors in parentheses. $^{***}p < 0.01$, $^{*}p < 0.10$. The full model includes 46 rating, 26 sectors, and 40 calendar-quarter fixed effects (coefficients omitted for brevity)

Column (1) of Table 3.3 reports the baseline results. The estimated greenium is $\hat{\beta} = -14.6$ basis points, statistically significant at the 10% level, suggesting that green-labelled bonds are priced modestly tighter than their conventional counterparts. However, once the regression includes rating, sector, and calendar-quarter fixed effects in column (2), the greenium estimate shrinks substantially to –6.7 basis points and becomes statistically insignificant ($p \approx 0.29$). These findings imply that observable issuer and market characteristics explain most of the yield differential previously attributed to the green label alone.

Scatter plots of yield versus maturity and tranche size corroborate the regression results: Conventional risk premia associated with bond tenor and issue size dominate the cross-sectional variation in yields (Fig. 3.5).

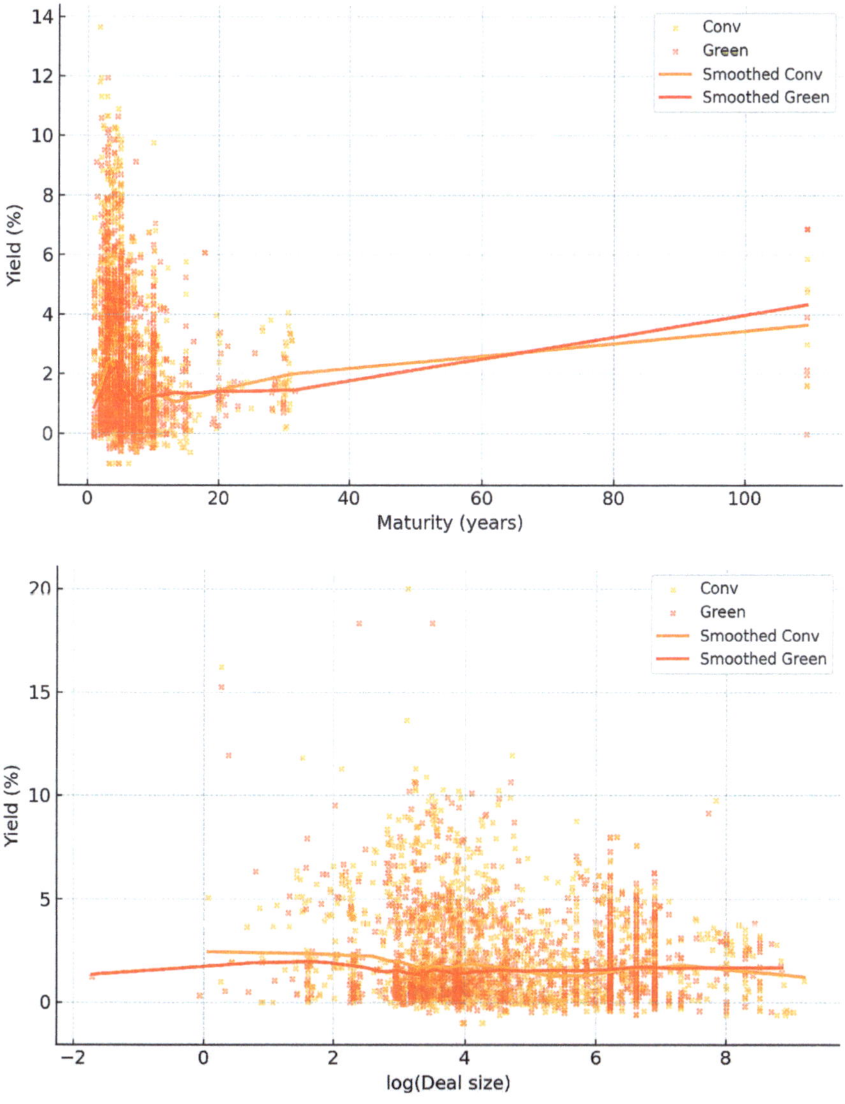

Fig. 3.5 Yield vs. maturity (left) and yield vs. size (right)

Two structural forces may explain the disappearance of a statistically robust greenium:

1. **Scarcity arbitrage**. The volume of green issuance has quadrupled since 2017. As supply grows to meet demand, excess price pressure on green bonds dissipates, eroding any scarcity-induced premium.
2. **Transparency arbitrage**. Investors increasingly discount labels that are not backed by rigorous ex post reporting and verification mechanisms; see Di Tommaso et al. (2025).

What institutional mechanisms might restore a pricing advantage for green bonds?

1. **Regulatory stringency**. The EU Green Bond Regulation (Reg. EU 2023/2631) introduces mandatory third-party verification and enforces alignment with the EU Taxonomy. This policy bundle enhances credibility and may re-establish pricing benefits via reduced informational asymmetry.
2. **Monetary policy channels**. If central banks were to assign preferential collateral treatment—e.g. via lower haircuts—to green instruments, this could translate into repo funding advantages and, by arbitrage, into primary-market pricing benefits.

Analysing a decade-long panel of 3326 euro-denominated primary-market bond tranches, we find no statistically significant greenium once standard controls are included. This result is robust across rating strata and coupon structures. The green label, in its current form, no longer signals scarcity nor conveys sufficient credibility to compress yields at issuance. Only a comprehensive policy framework—comprising credible labelling standards, stringent disclosure regimes, and preferential monetary treatment—can sustainably re-anchor green bond yields below their conventional counterparts.

3.4 Final Reflections

This chapter set out to examine whether green bonds are priced differently from their conventional counterparts—often referred to as "brown" bonds—and whether any such pricing differential constitutes a "greenium". Drawing on a comprehensive dataset of euro-denominated primary-market issuances, we tested whether green labels still confer a measurable cost-of-capital advantage in today's deep, information-rich bond markets.

Our empirical analysis reveals that once key structural features such as credit rating, issuance size, sector, and timing are accounted for, green bonds no longer exhibit a statistically significant yield discount relative to comparable conventional bonds. This marks a departure from earlier studies such as Zerbib (2019) and Baker et al. (2023), which had documented greenium effects ranging from 2 to 35 basis points in earlier phases of the market. While some subgroups—such as sovereign or highly rated issuers—have historically exhibited modest pricing advantages, particularly in earlier market periods (Ehlers and Packer 2017; Zerbib 2019), our results suggest that these benefits have largely dissipated by 2024.

The decline of the greenium appears driven by two structural forces: (i) the erosion of scarcity value as green bond issuance has scaled dramatically, and (ii) the waning informational content of the green label in the absence of rigorous post-issuance verification. Consistent with Tang and Zhang (2020), we find that labels without credible certification or detailed use-of-proceeds disclosures no longer serve as reliable signals to investors. Market participants now demand stronger institutional guarantees, and green bonds that fail to meet these evolving expectations are increasingly priced on par with conventional instruments.

These dynamics highlight the fact that green bond pricing is shaped not by labelling alone but by a more complex interplay of credibility, transparency, and regulatory context. As Kapraun and Scheins (2019) argue, yield differentials in sustainable debt markets are sensitive to sectoral risk, liquidity conditions, and investor perception—none of which can be fully captured by a binary green designation.

Looking ahead, the policy landscape may play a decisive role in re-establishing pricing differentials. The implementation of the European Green Bond Standard (Regulation EU 2023/2631) introduces mandatory taxonomy alignment and third-party verification, potentially restoring trust and pricing benefits to compliant issuers. Additional mechanisms—such as preferential treatment in collateral frameworks or supervisory incentives—could further enhance the cost effectiveness of green finance. Principles-based frameworks like the ICMA Green Bond Principles (ICMA 2024) remain relevant but may need greater enforcement power to influence pricing at scale.

In conclusion, green bonds continue to serve as important vehicles in the sustainable finance architecture, but their effectiveness in lowering borrowing costs now depends on institutional quality rather than symbolic labelling. Future research should further investigate how regulatory design, monetary incentives, and investor behaviour intersect to shape the evolving economics of labelled debt markets.

References

M. Baker, D. Bergstresser, G. Serafeim, J. Wurgler, The pricing and ownership of U.S. green bonds. Annu. Rev. Financ. Econ. **14**(1), 415–437 (2023)

J. Caramichael, A.C. Rapp, The green corporate bond issuance premium. J. Bank. Financ. **162**, 107126 (2024)

A.F. Cicchiello, M. Cotugno, S. Monferrà, S. Perdichizzi, Credit spreads in the European green bond market: a daily analysis of the covid-19 pandemic impact. J. Int. Financ. Manag. Account. **33**(3), 383–411 (2022a)

A.F. Cicchiello, M. Cotugno, S. Monferrà, S. Perdichizzi, Which are the factors influencing green bonds issuance? Evidence from the European bonds market. Financ. Res. Lett. **50**, 103190 (2022b)

Climate Bonds Initiative, Climate bonds standard. globally recognised, Paris-aligned certification of debt instruments, entities and assets using robust, science-based methodologies (2024). Accessed 12 June 2025
C. Di Tommaso, S. Perdichizzi, S. Vigne, A. Zaghini, Is the government always greener? Working-paper version on EconStor (2025)
M. Dutordoir, S. Li, J.Q.F. Neto, Issuer motivations for corporate green bond offerings. Br. J. Manag. **35**(2), 952–973 (2024)
T. Ehlers, F. Packer, Green bond finance and certification. BIS Q. Rev. September (2017)
S. Fatica, R. Panzica, M. Rancan, The pricing of green bonds: are financial institutions special? J. Financ. Stability **54**, 100873 (2021)
W. Febi, D. Schäfer, A. Stephan, C. Sun, The impact of liquidity risk on the yield spread of green bonds. Financ. Res. Lett. **27**, 53–59 (2018)
C. Flammer, Corporate green bonds. J. Financ. Econ. **142**(2), 499–516 (2021)
H.A. Hacıömeroğlu, S. Danışoğlu, Z.N. Güner, The grass is greener on the other side: comparison of green versus brown corporate bonds. Borsa Istanbul Rev. **22**(6), 1182–1194 (2022)
Y. Han, J. Li, Should investors include green bonds in their portfolios? Evidence for the USA and Europe. Int. Rev. Financ. Anal. **80**, 101998 (2022)
X. Hu, A. Zhong, Y. Cao, Greenium in the Chinese corporate bond market. Emer. Markets Rev. **53**, 100946 (2022)
X. Hu, B. Zhu, R. Lin, X. Li, L. Zeng, S. Zhou, How does greenness translate into greenium? Evidence from China's green bonds. Energy Econ. **133**, 107511 (2024)
C.Y. Huang, D. Dekker, D. Christopoulos, Rethinking greenium: a quadratic function of yield spread. Finance Res. Lett. **54**, 103710 (2023)
ICMA, Green bond principles. voluntary process guidelines for issuing green bonds (2021). Accessed 12 June 2025
ICMA, Sustainability-linked bond principles (2024). Accessed 28 May 2025
J. Kapraun, C. Scheins, Incredibly green: which bonds trade at a green premium (2019). Working Paper Goethe University Frankfurt, Germany
C. Koziol, J. Proelss, P. Roßmann, D. Schweizer, The price of being green. Financ. Res. Lett. **50**, 103285 (2022)
D.F. Larcker, E.M. Watts, Where's the greenium? J. Account. Econ. **69**(2–3), 101312 (2020)
Q. Li, K. Zhang, L. Wang, Where's the green bond premium? Evidence from China. Financ. Res. Lett. **48**, 102950 (2022)
K.U. Löffler, A. Petreski, A. Stephan, Drivers of green bond issuance and new evidence on the "greenium". Eurasian Econ. Rev. **11**(1), 1–24 (2021)
S. MacAskill, E. Roca, B. Liu, R.A. Stewart, O. Sahin, Is there a green premium in the green bond market? Systematic literature review revealing premium determinants. J. Cleaner Prod. **280**, 124491 (2021)
M.A. Naeem, O.B. Adekoya, J.A. Oliyide, Asymmetric spillovers between green bonds and commodities. J. Cleaner Prod. **314**, 128100 (2021)
A. Okafor, M. Adusei, O.C. Edo, Effects of the utilization of green bonds proceeds on green bond premium. J. Cleaner Prod. **469**, 143131 (2024)
L. Pham, Is it risky to go green? A volatility analysis of the green bond market. J. Sustain. Financ. Invest. **6**(4), 263–291 (2016)
J.C. Reboredo, Green bond and financial markets: co-movement, diversification and price spillover effects. Energy Econ. **74**, 38–50 (2018)
I. Sangiorgi, L. Schopohl, Why do institutional investors buy green bonds: evidence from a survey of European asset managers. Int. Rev. Financ. Anal. **75**, 101738 (2021)
D.Y. Tang, Y. Zhang, Do shareholders benefit from green bonds? J. Financ. Econ. **135**(2), 635–653 (2020)
E. Teti, I. Baraglia, M. Dallocchio, G. Mariani, The green bonds: empirical evidence and implications for sustainability. J. Cleaner Prod. **366**, 132784 (2022)
C. Tolliver, A.R. Keeley, S. Managi, Green bonds for the Paris agreement and sustainable development goals. Environ. Res. Lett. **14**(6), 064009 (2019)

Y. Wu, Are green bonds priced lower than their conventional peers? Emer. Markets Rev. **52**, 100909 (2022)

M. Yan, X. Li, X. Zhao, Z. He, Liquidity and cost advantage of green bonds. Financ. Res. Lett. **71**, 106433 (2025)

O.D. Zerbib, The effect of pro-environmental preferences on bond prices: evidence from green bonds. J. Bank. Financ. **98**, 39–60 (2019)

Chapter 4
Conclusion

Abstract This concluding chapter synthesises the book's central claims. First, architecture dominates instruments: Green bonds mobilise long-duration capital only when embedded in credible taxonomies, independent verification, and standardised ex post impact reporting. Second, heterogeneity is first order: issuer type, sector, maturity, and rating shape issuance and pricing; willingness to pay depends on verifiability and measurable outcomes, not labels. Third, market development is policy-elastic: clear rules, supervisory guidance, and transparency mandates crowd in demand, deepen liquidity, and improve data quality. Empirically, the "greenium" is conditional rather than universal and largely hinges on disclosure credibility and impacts measurement discipline. Operationally, the lesson is to signal with evidence—machine-readable allocation and impact metrics, consistent templates, and audit trails that lower verification costs and keep any premium tethered to real outcomes. Looking forward, artificial intelligence can strengthen sustainable finance across risk assessment, monitoring, and assurance; combined with distributed ledgers, it supports traceability and automated covenant checks. These gains require guardrails: governance by design, data stewardship, scaled assurance, and traceability. Policy should pair standards with teeth (mandatory post-issuance reporting and independent assurance) with digital public goods (open taxonomies, reference datasets, impact APIs) and incentives linked to eligibility and collateral treatment. The destination is a high-credibility equilibrium that aligns private capital with a just and resilient transition.

Keywords Sustainable finance · Green bonds · Greenium · Impact reporting · Monitoring-reporting-verification (MRV) · Verification and assurance · Artificial intelligence · Blockchain and traceability · Taxonomy and standards · Collateral policy and incentives

A. F. Cicchiello, S. Perdichizzi, *The Role of Green Bonds in Finance*, SpringerBriefs in Finance, https://doi.org/10.1007/978-3-032-14840-7_4

4.1 Green Bonds in Retrospect: Key Findings and Lessons Learned

A credible path to sustainability requires more than incremental policy tweaks or isolated technological breakthroughs; it hinges on how societies price externalities, allocate risk, and govern information. Green bonds have been pivotal in this reframing by coupling environmental objectives with investable fixed-income structures at scale (Maltais and Nykvist 2020; Bhutta et al. 2022). This book integrated conceptual analysis with tranche-level evidence to assess where green bonds deliver, where they fall short, and which design elements matter most.

Three conclusions stand out. *First*, architecture dominates instruments. Labels perform only as well as the institutions around them—credible taxonomies, robust second-party or certified verification, and standardised ex post impact reporting. Where these are weak, price and allocation patterns converge towards conventional debt and investor trust decays (Maltais and Nykvist 2020; Bhutta et al. 2022). *Second*, heterogeneity is first order. Issuer type, sector, maturity, and rating shape both issuance and pricing; investors rationally condition any willingness to pay on verifiability, alignment with material goals, and measurability of outcomes rather than branding per se. *Third*, market development is policy-elastic. Clear rules, supervisory guidance, and transparency mandates crowd in demand and scale supply; where institutional clarity improves, secondary-market resilience and data quality follow.

Empirically, we revisited whether green bonds are priced differently from otherwise similar bonds. Our evidence shows that any "greenium" is conditional, not universal. It appears in specific regimes and is sensitive to issuer credibility, instrument design, and macro-financial variables such as interest-rate cycles and capital-account conditions. Most importantly, disclosure quality and impact measurement discipline systematically predict investor readiness to pay for the label. These results are consistent with a broader theme of the book: In sustainable finance, *governance of information* is capital formation.

The operational lesson is straightforward. Signal with evidence, not adjectives. Ex ante eligibility must be matched by ex post allocation and impact metrics that are machine-readable, comparable across issuers and time, and auditable. Investors price the covenant, not the colour. Standard data templates and assurance processes reduce verification costs, sustain liquidity where deserved, and keep the premium—when it exists—tied to real impact rather than to scarcity or marketing.

Green bonds are neither a panacea nor a placebo. They are a powerful conduit *when* embedded in a high-credibility equilibrium of standards, verification, and transparent impact reporting. Within that equilibrium, they mobilise long-duration capital towards projects with measurable environmental and social benefits (Maltais and Nykvist 2020; Bhutta et al. 2022).

Two additional lessons deserve emphasis. The first concerns the interaction between green bond design and the broader monetary-financial environment. Price advantages attributed to colour compress or widen with liquidity regimes, benchmark eligibility, and collateral policy. When high-grade labelled paper is eligible in central-bank operations or purchase programmes, liquidity premia shrink and the marginal investor is more willing to finance long-duration projects; when eligibility wanes or policy uncertainty rises, any residual discount is the first to evaporate (Altavilla et al. 2021; Zaghini 2024). The implication is not that green premia are illusory, but that they are layered on conventional term, credit, and liquidity premia that remain dominant. Pricing claims that abstract from these layers will overstate the role of labels (Ehlers and Packer 2017).

The second concerns the distinction between allocation claims and impact claims. Use-of-proceeds covenants reallocate funds towards eligible assets and generate portfolio-level metrics; they do not, by themselves, guarantee real-world additionality. Additionality arises when labelled finance accelerates the timing or scale of investments relative to a credible counterfactual, or when it catalyses organisational change that would not otherwise occur. Distinguishing the two requires a data architecture that links tranche-level proceeds to project execution and operating outcomes under a transparent baseline. Where that architecture is present, the label helps coordinate expectations among issuers, reviewers, and investors; where it is absent, markets default to heuristics and price signals lose informational content (Flammer 2021; Zerbib 2019).

The policy translation is immediate. Credible taxonomy alignment, independent verification, and standardised ex post reporting are not compliance ornaments; they are production inputs for the information that fixed-income markets require to price risk and impact jointly. Regulatory initiatives that harden these inputs—minimum reporting fields, audit trails, and assurance quality thresholds—lower the unit cost of verification and raise the signal-to-noise ratio in secondary trading. Supervisory guidance can reinforce the equilibrium by clarifying how labelled exposures are treated in risk management and by linking eligibility, where appropriate, to disclosure depth rather than to labels alone. Market infrastructure can do the rest: machine-readable templates, open repositories of allocation and impact reports, and consistent identifiers that permit cross-issuer comparison and replication of empirical results.

Taken together, the accumulated evidence of this book points to a pragmatic equilibrium. Green bonds work best when they are treated as contracts for information and governance as much as for funding. They lower financing frictions for climate-aligned investment when issuers can demonstrate, and investors can audit, the path from proceeds to outcomes. They can coexist with performance-linked instruments and outcome-based contracts, each occupying the segment where its information content is sharpest. The task ahead is therefore less about multiplying labels and more about scaling the institutional machinery that turns those labels into credible, comparable, and decision-useful evidence. In that setting, the contribution of green bonds is an empirical matter: Enhancements in information and assurance regimes should precede and predict higher investment and verifiable impact, conditional on risk.

4.2 The Future of Green Finance and the Role of Artificial Intelligence

Artificial intelligence (AI) is poised to reshape sustainable finance across the information lifecycle: data generation, curation, inference, and assurance. In risk assessment, fraud detection, impact evaluation, and forecasting, AI can improve precision, timeliness, and scale, enabling more resilient capital allocation (Hassanein and Tharwat 2024; Hassanien et al. 2024; Musleh Al-Sartawi et al. 2022). Machine learning and natural language processing already enhance ESG signal extraction, dynamic credit assessment, and real-time monitoring of financed projects (Rani and Singh 2024; Nepal et al. 2025). As a practical matter, these tools shift sustainable finance from claims to evidence by tightening feedback loops between project performance and investor oversight (Akhtar et al. 2025).

Two complementary enablers are salient. *First*, the diffusion of ICT in European banking has produced organisational blueprints—data pipelines, governance routines, and operational resilience—that are directly applicable to the digitalisation of sustainable finance (Cicchiello et al. 2021). *Second*, distributed ledgers can harden auditability. When combined with AI, blockchain supports traceable use of proceeds, automated covenant checks, and tamper-evident impact reporting, thereby lowering verification costs and reducing information asymmetries (Dorfleitner and Braun 2019; Kulkov et al. 2024). These capabilities target greenwashing at its roots by standardising evidence chains and enabling anomaly detection at scale (Kumar et al. 2025; Zhao et al. 2024).

AI can also democratise participation. Automated advice and personalisation can tailor green portfolios for retail investors; digital issuance and tokenisation can broaden access to labelled debt and adjacent ESG products while lowering search and transaction costs (Wang et al. 2025; Maltais and Nykvist 2020). If paired with robust assurance, wider participation deepens demand, supports liquidity, and stabilises flows into high-integrity assets.

These gains require guardrails. AI introduces ethical and systemic risks: bias propagation, privacy breaches, opacity of complex models, and new operational interdependencies (Hassanein and Tharwat 2024). A credible pathway rests on four pillars: (*i*) *governance by design*, with documentation, auditability, and human-in-the-loop controls for material decisions (Akhtar et al. 2025); (*ii*) *data stewardship*, including provenance tracking and consent management; (*iii*) *assurance at scale*, with fairness and robustness testing integrated into financial audit cycles (Kumar et al. 2025); and (*iv*) *traceability*, via on-chain attestations for use of proceeds and impact claims where appropriate (Dorfleitner and Braun 2019). With these in place, AI can amplify the effectiveness of green finance rather than merely accelerate its errors.

A finance-theoretic way to see these gains is through the cost of information production and the precision of measurement. In delegated-monitoring settings, the hurdle for financing a project with uncertain payoffs is a function of both the lender's verification cost and the variance of the signals used to underwrite and monitor performance. Applied to sustainable finance, learning systems that extract structured variables from heterogeneous disclosures, images, sensor feeds, and text reduce fixed appraisal costs and improve signal precision; both effects tighten confidence bands around pro-forma cash flows and impact metrics, which lowers required returns for otherwise bankable projects. In primary markets, better measurement curbs adverse selection by making it harder for low-quality issuers to mimic high-quality disclosure at low cost. In secondary markets, anomaly detection on allocation and impact series raises the expected penalty for misreporting, reinforcing the deterrence channel documented in studies of disclosure stringency (Cicchiello et al. 2021; Akhtar et al. 2025).

These efficiency gains matter only if they survive contact with institutional constraints. Model governance is therefore central. In credit and impact applications, out-of-sample stability and robustness to representation shift must be established before outputs enter covenant checks or risk limits. Backtesting against human benchmarks, falsification tests on synthetic counterfactuals, and drift diagnostics become part of the audit trail so that supervisors and investors can replicate assurance claims (Kumar et al. 2025). Where distributed ledgers are used, model artefacts—architectures, parameter hashes, validation summaries—can be time-stamped alongside use of proceeds and ex post metrics, creating tamper-evident links from raw data to reported outcomes (Dorfleitner and Braun 2019; Kulkov et al. 2024). The informational value of these links is not intrinsic to the technology but to the discipline it imposes: provenance, version control, and reproducibility.

A second design question concerns econometric identification. If AI-enabled verification coincides with rising investor demand, attributing price or volume changes to the technology rather than to preferences is non-trivial. Stepped-wedge deployments across issuers or portfolios, combined with difference-in-differences on pricing and allocation outcomes, provide implementable identification strategies. On the real-economy margin, the relevant test is whether projects screened and monitored with AI advance faster along construction and commissioning milestones and whether deviations from budget and schedule compress, conditional on sector and geography (Hassanein and Tharwat 2024; Rani and Singh 2024). These are measurable propositions and align with the shift from claims to evidence noted above (Akhtar et al. 2025).

Democratisation is a further margin where technology and market design interact. Recommendation engines and conversational interfaces can broaden participation by lowering search and advice costs for households and SMEs, while tokenised issuance reduces minimum lot sizes and settlement frictions (Wang et al. 2025; Maltais and Nykvist 2020). The welfare effect is positive only if suitability and fairness constraints travel with automation. Embedding consent management, explainability at the point of advice, and audit rights for clients is therefore not an ethical add-on but a condition for durable scale (Hassanein and Tharwat 2024). Finally, the sector must internalise the resources that AI itself consumes. Model selection, inference frequency, and data retention policies should be part of the environmental control framework so that the intensity of computation is proportionate to decision value. In a mature architecture—data pipelines with stewardship, models with governed uncertainty, ledgers that memorialise evidence—AI does not replace judgement; it makes the evidentiary chain cheaper, faster, and harder to game, thereby aligning capital with verifiable environmental outcomes (Zhao et al. 2024; Nepal et al. 2025; Musleh Al-Sartawi et al. 2022).

4.3 Final Reflections

The economics of labelled debt has moved from rhetoric to verification. In that shift, the central organising fact is that markets price information and the governance that produces it. Where eligibility, monitoring, and impact reporting are credible and comparable, labelled instruments reduce financing frictions for climate-aligned investment and transmit information about environmental performance through prices; where these preconditions are absent, labels attenuate to marketing and spreads revert to conventional risk premia (Maltais and Nykvist 2020; Bhutta et al. 2022). The book's evidence shows that instrument form is secondary to institutional quality: Investors do not pay for adjectives, and they pay for auditable covenants.

A first implication is outcome primacy. Financial terms and disclosures should be tethered explicitly to outcome metrics, with uncertainty acknowledged rather than hidden. In fixed income, that means use of proceeds mapped to measurable assets and services, allocation confirmed at completion, and impact reported in machine-readable form with clear baselines and error bounds (Maltais and Nykvist 2020). The purpose is not rhetorical hygiene but contractability: Only outcomes that can be measured and assured at reasonable cost can sustain a price signal for impact. The empirical patterns documented here are consistent with this logic. Pricing advantages for labelled paper are conditional, regime-dependent, and largest when verification quality is high; they compress when liquidity regimes shift or eligibility fades, revealing the layered structure of term, credit, and liquidity premia beneath (Ehlers and Packer 2017; Altavilla et al. 2021; Zaghini 2024).

A second implication is that credibility is a joint product. Issuers, reviewers, and investors must coordinate on standardised evidence chains in which data lineage, model documentation, and assurance thresholds are first-class objects. The technology now exists to support this compact at scale. Automated extraction and validation can surface inconsistencies and anomalous narratives across allocation and impact reports; anomaly detection raises the expected penalty for misreporting and reduces the option value of vague claims (Kumar et al. 2025; Zhao et al. 2024). Yet technology does not substitute for governance. Models must be reproducible, their drift monitored, and their role in decision processes auditable, or they risk accelerating error rather than elevating evidence (Hassanein and Tharwat 2024).

Inclusion matters for scale. If sustainable finance is to mobilise household and SME balance sheets as well as institutional ones, access frictions must fall without diluting integrity. Digital advice and personalisation can broaden participation, and digital issuance can reduce minimum lot sizes and settlement frictions; but suitability, privacy, fairness, and contestability protections must travel with automation end to end (Wang et al. 2025; Hassanein and Tharwat 2024). The relevant outcome is not click-throughs but informed allocation: Investors should be able to see, at acceptable cost, how their capital maps to assets, how those assets produce environmental services, and how uncertainty is treated in that mapping.

Public policy can accelerate convergence to a high-credibility equilibrium. Standards with teeth—mandatory post-issuance impact reporting and independent assurance—ought to be paired with digital public goods: open taxonomies, reference datasets, and APIs for allocation and impact metrics that lower verification costs and crowd in private analytics (Akhtar et al. 2025). Incentives should then reinforce integrity by linking eligibility, index inclusion, and collateral treatment to the depth and quality of disclosure rather than to labels alone (Maltais and Nykvist 2020). In such an architecture, the marginal dollar of compliance expands the stock of decision-useful information, improving allocative efficiency and accountability simultaneously.

The broader market design is complementary rather than zero sum. Use-of-proceeds bonds contract for evidence at the asset level; performance-linked instruments contract for issuer-level trajectories; outcome-based contracts tie repayments to measured services. Each occupies the segment where its information content is sharpest. In deep primary markets, credible labelling reduces adverse selection by making it harder for low-quality issuers to mimic high-quality disclosure at low cost. In secondary markets, machine-readable reports and consistent identifiers enable replication, factor-adjusted attribution, and enforcement of claims. Where these ingredients are present, the informational channel that underlies any green premium is visible and testable; where they are not, premium narratives collapse into selection and liquidity effects (Flammer 2021; Zerbib 2019).

The same logic extends to the interface with artificial intelligence. AI is not an end, but a means to compress the cost of information production and to raise the precision of measurement in delegated-monitoring settings. Learning systems can map heterogeneous disclosures to structured variables, align sensor and project data with covenants, and make evidence chains tamper-evident when coupled with traceability tools (Kumar et al. 2025; Dorfleitner and Braun 2019; Kulkov et al. 2024). These capabilities tighten feedback loops between project performance and investor oversight and move sustainable finance from claims to evidence (Akhtar et al. 2025). They are powerful only when embedded in institutions that privilege transparency over opacity and reproducibility over bespoke models.

The practical test for the next phase is therefore straightforward. Can we trace an auditable line from proceeds to outcomes, at portfolio and project level, with uncertainty accounted for and model risk governed? If the answer is yes, labelled debt functions as a contract for information as much as for funding, and the market's price signals will support a just and resilient transition. If the answer is no, premium stories will continue to be fragile, and the allocative promise of sustainable finance will remain contingent. Green bonds will evolve—more dynamic in structure, more data-rich in reporting, more digitally native in assurance—but their effectiveness will continue to depend on the credibility of the institutions that sit behind the label. The task is not to multiply labels; it is to scale the machinery that turns those labels into comparable, decision-useful evidence. That is the route by which finance serves, rather than sidesteps, the demands of the transition.

References

S. Akhtar, M. Alam, N.U.H. Wani, S.H. Jafar, *Green Horizons* (Springer, Berlin, 2025)

C. Altavilla, G. Carboni, R. Motto, Asset purchase programs and financial markets: lessons from the euro area. Int. J. Central Bank. **17**(5), 1–48 (2021)

U.S. Bhutta, A. Tariq, M. Farrukh, A. Raza, M.K. Iqbal, Green bonds for sustainable development: review of literature on development and impact of green bonds. Technol. Forecast. Soc. Change **175**, 121378 (2022)

A.F. Cicchiello, M. Cotugno, S. Monferra, S. Perdichizzi, Exploring the impact of ICT diffusion in the European banking industry: evidence in the pre- and post-COVID-19. J. Financ. Manag. Markets Instit. **9**(2), 2150010 (2021)

G. Dorfleitner, D. Braun, Fintech, digitalization and blockchain: possible applications for green finance, in *The Rise of Green Finance in Europe*, ed. by M. Migliorelli, P. Dessertine. Palgrave Studies in Impact Finance. (Palgrave Macmillan, Cham, 2019), pp. 207–237

T. Ehlers, F. Packer, The growing market for green bonds: finance and certification. BIS Q. Rev. (2017)

C. Flammer, Corporate green bonds. J. Financ. Econ. **142**(2), 499–516 (2021)

A. Hassanein, H. Tharwat, Artificial intelligence (AI) and green finance. in *The Palgrave Handbook of Green Finance for Sustainable Development* (Springer, Cham, 2024), pp. 859–889

A.E. Hassanien, A. Darwish, S. Elghamrawy (eds.), *Artificial Intelligence for Environmental Sustainability and Green Initiatives* (Springer, Berlin, 2024)

I. Kulkov, J. Kulkova, R. Rohrbeck, L. Menvielle, V. Kaartemo, H. Makkonen, Artificial intelligence-driven sustainable development: examining organizational, technical, and processing approaches to achieving global goals. Sustain. Dev. **32**(3), 2253–2267 (2024)

S. Kumar, D. Sharma, S. Rao, W.M. Lim, S.K. Mangla, Past, present, and future of sustainable finance: insights from big data analytics through machine learning of scholarly research. Ann. Oper. Res. **345**(2), 1061–1104 (2025)

A. Maltais, B. Nykvist, Understanding the role of green bonds in advancing sustainability. J. Sustain. Financ. Invest. 1–20 (2020). https://doi.org/10.1080/20430795.2020.1724864 Advance online publication

A.M. Musleh Al-Sartawi, K. Hussainey, A. Razzaque, The role of artificial intelligence in sustainable finance. J. Sustain. Financ. Invest. 1–6 (2022). https://doi.org/10.1080/20430795.2022.2057405. Advance online publication

R. Nepal, X. Zhao, K. Dong, J. Wang, A. Sharif, Can artificial intelligence technology innovation boost energy resilience? The role of green finance. Energy Econ. **142**, 108159 (2025)

M. Rani, R. Singh, Role of artificial intelligence in sustainable finance, in *Engineering Applications of Artificial Intelligence* (Springer, Cham, 2024), pp. 409–419

Q. Wang, T. Sun, R. Li, Does artificial intelligence (AI) enhance green economy efficiency? The role of green finance, trade openness, and R&D investment. Hum. Soc. Sci. Commun. **12**(1), 1–22 (2025)

A. Zaghini, Unconventional green. J. Corp. Financ. **85**, 102556 (2024)

O.D. Zerbib, The effect of pro-environmental preferences on bond prices: evidence from green bonds. J. Bank. Financ. **98**, 39–60 (2019)

C. Zhao, K. Dong, K. Wang, R. Nepal, How does artificial intelligence promote renewable energy development? The role of climate finance. Energy Econ. **133**, 107493 (2024)

MIX
Papier aus verantwortungsvollen Quellen
Paper from responsible sources
FSC® C105338

If you have any concerns about our products,
you can contact us on
ProductSafety@springernature.com

In case Publisher is established outside the EU,
the EU authorized representative is:
Springer Nature Customer Service Center GmbH
Europaplatz 3, 69115 Heidelberg, Germany

Printed by Libri Plureos GmbH
in Hamburg, Germany